A PHILOSOPHER LOOKS AT THE WEATHER

From barometers to the famous BBC shipping forecast, we have – over the centuries – developed the means to predict, harness, and shield ourselves from what is happening in the atmosphere. Attitudes about the planet's weather, as well as about human identity, have thereby taken on new meanings. In an era of climatic anxiety, what weather is and how weather behaves have taken on additional currency. Benjamin Hale weaves together philosophy and anecdote into a many-faceted exploration of this powerful force that shapes who we are and how we think about our place in the world. He argues that in our drive to "scientize" weather, with all the technological advances in managing, anticipating, and understanding it, we also risk distancing ourselves from weather and losing a complete sense of what it is. This entertaining book reminds us that the weather is and always will be in some sense outside our control, and that consequently we are and forever will be learning to live alongside it.

BENJAMIN HALE is Professor in the Departments of Environmental Studies and Philosophy at the University of Colorado, Boulder. He is the author of *The Wild and the Wicked: On Nature and Human Nature* (2016), co-editor of the journal *Ethics, Policy and Environment*, and former President of the International Society for Environmental Ethics.

A Philosopher Looks at

In this series, philosophers offer a personal and philosophical exploration of a topic of general interest.

Books in the series

Nancy Cartwright, *A Philosopher Looks at Science*
Sophie Grace Chappell, *A Philosopher Looks at Friendship*
Raymond Geuss, *A Philosopher Looks at Work*
Paul Guyer, *A Philosopher Looks at Architecture*
Benjamin Hale, *A Philosopher Looks at the Weather*
Zena Hitz, *A Philosopher Looks at the Religious Life*
Kate Moran, *A Philosopher Looks at Clothes*
Stephen Mumford, *A Philosopher Looks at Sport*
Onora O'Neill, *A Philosopher Looks at Digital Communication*
Michael Ruse, *A Philosopher Looks at Human Beings*

A PHILOSOPHER LOOKS AT
THE WEATHER

BENJAMIN HALE

CAMBRIDGE
UNIVERSITY PRESS

Shaftesbury Road, Cambridge CB2 8EA, United Kingdom

One Liberty Plaza, 20th Floor, New York, NY 10006, USA

477 Williamstown Road, Port Melbourne, VIC 3207, Australia

314–321, 3rd Floor, Plot 3, Splendor Forum, Jasola District Centre,
New Delhi – 110025, India

103 Penang Road, #05-06/07, Visioncrest Commercial, Singapore 238467

Cambridge University Press is part of Cambridge University Press & Assessment, a department of the University of Cambridge.

We share the University's mission to contribute to society through the pursuit of education, learning and research at the highest international levels of excellence.

www.cambridge.org
Information on this title: www.cambridge.org/9781009402682

DOI: 10.1017/9781009402675

© Benjamin Hale 2026

This publication is in copyright. Subject to statutory exception and to the provisions of relevant collective licensing agreements, no reproduction of any part may take place without the written permission of Cambridge University Press & Assessment.

When citing this work, please include a reference to the DOI 10.1017/9781009402675

First published 2026

Cover image: golovorez / E+ / Getty Images

A catalogue record for this publication is available from the British Library

A Cataloging-in-Publication data record for this book is available from the Library of Congress

ISBN 978-1-009-40268-2 Paperback

Cambridge University Press & Assessment has no responsibility for the persistence or accuracy of URLs for external or third-party internet websites referred to in this publication and does not guarantee that any content on such websites is, or will remain, accurate or appropriate.

For EU product safety concerns, contact us at Calle de José Abascal, 56, 1°, 28003 Madrid, Spain, or email eugpsr@cambridge.org

CONTENTS

Introduction 1

1   Neither Rain, nor Snow, nor Sleet, nor Hail: What Is Weather? 4
    A Disquisition on Definition 20
    Weather in Mythology, Literature, and Language 24
    A Final Stab at a Definition 36

2   Of Rain and Parades: How Weather Affects Us 40
    General Observations about Impacts and Effects 42
    Quantitative Metrics: How Much It Impacts Us 52
    Qualitative Effects: How It Shapes Us 63
    Environmental Determinism 69

3   Battening the Hatches: Living with Weather 80
    Dispositional Orientations toward the Weather 82
        Accepting It 84
        Resisting It 90
        Harnessing It 97
    The Social Construction of Weather: Facts, Hazards, and Boons 108

4   Seeing through the Fog: Predicting the Weather 118
    Meteorology in Antiquity 120
    Renaissance and Enlightenment Meteorology 128
    Modern Meteorology 136
    Building the Conceptual Apparatus 140

From Ideas to Words   145
From Words to Numbers   149

5   Catching a Cloud: Controlling the Weather   159
   Rainmakers: Influencing the Gods   165
      The Teleological Model   170
      Mechanistic and Other Models   175
   Pluviculturists: Pulling the Knobs and Levers   178
      Reintroducing Intent   183
   The Temptation of Control and the Folly
   of Power   188

Conclusion: Silver Linings   203

Notes   210
Bibliography   231
Index   243

Introduction

It is a common bromide and accepted truism that if one has nothing to talk about, then one can always talk about the weather. From office parties to high school reunions, from blind dates to cross-Atlantic airline flights, weather is the go-to conversation starter that rarely succeeds in starting the conversation. That makes it particularly strange that a philosopher, a person who generally has too much to say – indeed, who belongs to a class of intellectuals deemed so stuffy and smug as to pride themselves entirely on the alleged depth and meaning of the things they say – would stoop so low as to talk about the weather.

And yet, here we are.

Weather is a fair bit more complicated than it first seems. It is both essential for life on earth and wildly hazardous to life on earth. It is arguable, but I think true, that the very idea of civilization is dependent upon weather, and yet we tend to think of it as an anti-civilizing force, the kind of thing that comes along and lays ruin to our homes and villages. It is arguable, but I think also true, that the very idea of humanity is dependent on weather, and yet we also tend to think of it as an anti-humanizing force, a kind of uninvited guest who drunkenly lays waste to our weddings. Most of our ramparts and walls have been built, if not to keep out dragons and marauders, then to buttress ourselves against the various hazards that weather throws our way. Most of our cultural festivals and family meals have been organized, if not only to keep us plump and entertained, then to protect nature's harvest against the corrosive effects of the weather. Though we do often talk about how weather has worked us over, what we don't often think about is how we, through our practical interventions and conceptual explorations, have helped shaped the weather as well.

This book is divided into five chapters, each relatively short and approachable. It is bookended by this Introduction and a slightly longer Conclusion. Each chapter builds on the previous chapter and seeks to make a claim about weather overall. In Chapter 1, I explore definitional questions and aim to get a grip on what we mean when we talk about weather. In Chapter 2, I turn to explore the ways in which weather affects us. In Chapter 3, I flip that discussion on its head and explore attitudes and orientations toward weather, aiming there to explore the ways in which our changing technological and political landscape has shaped our thinking about weather. In Chapter 4, I delve into questions about the origins of and suppositions implicit in weather prediction and meteorology. And in Chapter 5, I look at historical and contemporary efforts to control the weather.

The thesis I chase is that weather doesn't admit of a stable definition. It is not something that sits passively, specimen-like, on a shelf, but is instead an all-encompassing dynamic and volatile medium we live with and in. It is nothing so simple as a collection of discrete events – rain or sleet or hail – but instead a howling force that surrounds us, that seems to have a mind of its own, that works us over in ever-changing ways. It shapes who we are and how we live our lives. But just as weather is ever-changing, so too is our relationship with weather also ever-changing. Over the past several thousand years, given the innovations and interventions of civilization, weather has morphed from the tempestuous activity of the gods to an unceasing string of mostly mild inconveniences. These days, a large portion of the global population witnesses the world's most massive superstorms from the comfort of their living rooms, munching their dinner while thinking only abstractly of the distant hapless souls who have had their lives upended by a devastating hurricane or tornado. Weather has transformed from an unpredictable force shaping every aspect of our lives to a form of entertainment that we can swipe past on our smartphones. Only every once in a while do the inconveniences of weather roar massively and violently through our lives, but when they do, they can serve as a not-so-gentle reminder of our place in the cosmic scheme of things.

1   Neither Rain, nor Snow, nor Sleet, nor Hail
What Is Weather?

Picture an office party. There you are, standing alone near the punch bowl. Over comes Clive, that annoying guy from the cubicle on the second floor. "Hi," he says, eyeing the deviled eggs. "Hi," you reply. There is a moment of awkward silence. "Beautiful weather today," he says. "Yes," you nod. "They're saying rain tomorrow," he says again. "Yes," you reply, hoping to avoid further engagement. He takes an egg, lifts a napkin from the stack, and walks away.

That's it. That's the whole conversation.

Though many of us are conversant in matters of the weather and are likely aware of the basics – what's happening outside, the forecast for the week, how to interpret temperature, when clouds billow above us, what kinds of precipitation to expect when we see an umbrella, maybe even how major meteorological events in far-off lands are affecting populations we scarcely understand – most of us are at a loss when asked to say much more than that.

And that's why weather is as much a conversation starter as it is a conversation stopper.

If you really want to stop the conversation dead in its tracks, however, rather than following up with the expectable chit-chat about the forecast, pause for a moment, put a contemplative furrow in your brow, and ask your unwitting interlocutor not what the weather is going to be but rather what weather is.

Say something like this: "What exactly *is* weather?" You'll likely get looks of confusion, as though you are a moron. And then you may get a sideward glance, to gauge your sincerity. You may, after a moment, get an attempt at an answer. "It's what's happening outside." And then your interlocutor will likely walk away, assuming you to be one of those frustrating existentialist types they remember from college.

Weather is such a commonplace, so familiar to almost all of us, that the question borders on the absurd. Everyone knows what weather is. Don't be daft!

But since we are gamboling through the philosophical weeds here, it is precisely our task in this book to ask such questions. We aim – together – to think a little more deeply about weather; to cut through the fog and get clarity on what is otherwise a hazy notion. Our task in this book on philosophy and weather is not to answer the question about *how* the weather is, or *what* it's going to be, but really to understand weather, to wrap our minds around it; even, dare I say, to wrap our minds around how other minds have wrapped themselves around weather. To do this, it would make sense first to get a basic sense of what we're talking about, to come up with a rough-and-ready definition that we can use throughout this expedition. Unfortunately, that task, it turns out, is not as simple as tossing back a deviled egg and blasting out a definition.

Nevertheless, it's a pretty good place to start.

****

The simplest, most basic answer to the question of what weather is begins by listing off a litany of different examples. Weather is: rain, snow, sleet, hail, clouds, sun, wind, storms, cold snaps, heat waves, clear skies, rainbows, thunderbolts and lightning, very very frightening. Every one of us is familiar with events such as these. We've all had personal experiences with many types of weather and sometimes extreme weather events. No doubt, we also all have our own stories about the blizzard that defined our childhood, the hurricane that upended our travels, or the beating sun that torched our skin on vacation.

Our own personal experiences hardly exhaust all possible manifestations of weather, of course, as we are all also aware of at least some of the less common, albeit regularly occurring, weather events: tornadoes, hurricanes, supercells, thundersnow, dust storms, tsunamis, and so on. Once one begins thinking about more offbeat weather events such as these, it's not difficult to veer into much more exotic territory. You may have heard of waterspouts, which are essentially tornadoes that form over large bodies of water, but have you heard of fire whirls or fire tornadoes? Those exist too. These often occur in forest fires or bushfires, where the heat from the fire is so intense that it causes wind to swirl up in a vortex. Imagine the terror of looking over your shoulder only to see a fire tornado bearing down upon you!

Ever heard of a *jökulhlaup*? I hadn't either until I learned the term from an Icelandic woman thirty years ago. *Jökulhlaups* are glacial outburst floods, which occur when warming temperatures and glacial meltwater slowly melt away the lower portion of an existing ice wall, releasing torrents of water into the valley below, often destroying villages. While most have historically been associated with Iceland, *jökulhlaups* have been recorded on other continents as well.

Such exoticism need not be deadly, of course, but instead can provide fantasy fodder for Instagram and other social media platforms: Godlike crepuscular rays make for the most extraordinary images, but even before cameras they surely must've stricken awe and wonder even in skeptics about greater powers. Mammatus clouds, shaped like bubbles, gurgle and glide along the sky, providing a surreal blanket that envelops the scene. Frost flowers form flower-like ice sculptures on the ground that look to observers like a world frozen in time. Hair ice (also called ice wool or frost beard) is a somewhat rare phenomenon that occurs when fungus on tree bark causes ice crystals to form in fine strands of hair that look like eyebrows or wool sweaters on decaying logs or trees. Ball lightning hovers in the air or crawls across the ground as a ghostly dancing ball, looking to observers like a phantom visitor from another plane. Haboobs occur when thunderstorms fall in on themselves and stir up a massive cloud of dust and dirt, often in dry, desert environments, overwhelming the scene and blocking out the light. Verga storms, or "dry" storms, are unusual precipitation events with rain on days so hot that the rain evaporates before it hits the ground. Brinicles are underwater ice stalactites that, seen from below, look like tornadoes from an alien world. These exotic manifestations of weather seem almost magical in nature, calling to mind the air, water, and fire elementals common in Dungeons & Dragons. Imagine the thumbs up and likes you might get off an image of any of these!

The weirdness of weather doesn't stop there. There are notable events throughout history in which hundreds of animals – birds, frogs, fish, locusts, worms – fall from the sky like oversized hailstones pummeling the earth.[1] The phrase "it's raining cats and dogs" is common enough but usually not taken literally. Most of us will likely never witness an event such as this, as it is indeed highly unusual, or at least occurs in highly specific conditions that are hard for even seasoned meteorologists to understand. There are several hypotheses about what might account for these phenomena: that large clusters of animals are swept up by strong winds; that they don't actually fall at all, but are instead brushed along the ground by wind or flooding; or that there are tornadic waterspouts that lift them high into the sky and drop them.[2] Should incidents such as these be considered in our answer to the question of what weather is?

Even other events have the shape and feel of weather but don't so tidily fall into the category that most might otherwise consider to be weather. Volcanic lightning, for instance, occurs when the forces of the earth interact with electrically charged particles in the sky, essentially during volcanic eruptions. Is that weather? On one hand it would seem that it is not. Volcanic lightning is the consequence of a volcanic eruption, not the consequence of cloud formations and wind. On the other hand, in many respects it is not all that different from lightning during thunderstorms, which occur not because rock, ash, and ice have been emitted into the sky, but because a bunch of water has been stirred up by other conditions. Electrical charges are doing the same thing they do in the sky – seeking a pathway to get down to earth. What should it matter whether they're passing through water that evaporated from the heat of the beating sun or was instead pressurized through the blast furnace of a volcano? The problem with categorizing such events as "weather" is that similar such lightning-like charges appear from other physical interactions as well: ice charging, frictional charging, and fracto-emission (generation of a charge from fracturing rock). If volcanic lightning counts as weather, then it would seem that we also have to consider electrical charges that emerge from fractured rock as weather too.

Likewise, other anthropogenic phenomena also fall into the category of weather events. "Corn sweat" is sometimes reported as weather. Never heard of it? This is the superhumid environment created when the transpiration of corn crops causes additional humidity to fill the air. Are we to consider corn sweat weather? It feels uncomfortable to do so, but why should we not? Does it matter whether transpiration comes off a body of water or from the interior of a vegetable?

What we seek here is a definition for "weather"; and the word "weather," after all, is a substantive, a noun. It's not a verb or an adjective or an adverb. It is not an action or a description or a modifier. It is a substantive *thing* – a *real* thing – something each one of us grapples with every day. And inasmuch as it's a substan-*tive*, it is and implies substan-*ces*. It's much easier for us to understand common nouns when we can point to those things about which we're talking.

But to answer the question in this way is a bit like answering the question of what a dog is and getting back the response that a dog is a poodle or a Chihuahua or a golden retriever. Those are all *examples* of dogs, but not what a dog is. So it does not seem, at least on first analysis, that we can gain clarity on our inaugural question simply by looking at a list of examples.

****

A slightly different approach to the question of what weather is might triangulate toward an answer by generalizing from a known list of instances of weather and proposing that weather is not a thing itself but a class of things. When we say that there will be weather, we just mean more generally that there will be some kind of something falling from the sky, whether rain or snow or sleet or hail. Reformulating the definition this way makes sense, since precipitation is a thing, or a bunch of little droplet-like things, and when asked for a definition, it is natural to gravitate toward things. The classing of things is easier to point to than abstract ideas or concepts. Often weather comes in the form of substances falling from the sky – precipitates – rain, snow, sleet, hail. Wind, conceivably, if you think of wind as many molecules of air bunched together and pushing against us, is also a thing. It's only natural then to think of weather as the stuff that is thrown at us: water or ice from the sky. On this view, any precipitation is just a particular thing-like instantiation of a more general form. Weather is the *class* of things that captures all of them.

But weather, if it is a class of things, remains an elusive class, as the various instantiations of precipitation don't exhaust what most of us would count as weather. Though we can point to, touch, and walk through precipitation, in the same way that we might walk through gelatin or soup, we can't so easily point to cold or heat, which generally also fall in the catch-bin of weather.

Weather in this way is not just a *class* of different substances but maybe a class of phenomena. It includes, among other phenomena, light and heat, cloudiness and cold, rainbows and rays of sun, which aren't so easily classed as "things." Temperatures and colors and states of affairs are an essential part of this mix as well. How are we to understand temperatures? How are we to think of the interplay of light? Should these details factor into our understanding of weather? It seems that they should.

The problem grows all the more pronounced when we consider that weather is neither static nor inert. It changes all the time and is one of the most characteristically change-driven phenomena we experience. Where I live in Colorado people often say things like, "if you don't like the weather, wait an hour."[3] For some reason, most who utter this phrase believe that the phrase applies only to this region, but it turns out that people in Minnesota, Florida, New York, and probably everywhere else on the planet also have a similar saying. In a way it's more of a statement about weather than it is about the weather in a given location. That is: Weather is constantly changing. It does not sit statically in front of us like a rock or a mountain. We cannot ignore it like a blank wall or an empty backdrop. It is always moving, shifting shape, and changing around us to avoid capture and definition.

Moreover, the things included in the class of weather phenomena are rarely alike, and in some ways radically different. In the early years of planet Earth, 2.4 billion years ago, weather didn't manifest in the same way that it does now.[4] The skies were filled with different gases, and sometimes it would rain methane, leaving the ground shrouded in a sulfur fog. As a consequence, the Earth's sky back then would have looked pink or red or taken on a different hue.

Taking a longer view, current science suggests that three billion years ago the sun was considerably weaker than it is now, emitting approximately 20 percent less energy. The so-called faint young Sun paradox presents paleo-meteorologists with a conundrum: Given that the sun would have been too weak to warm the Earth to the point at which water would melt, rain back in the early days of the planet should have been impossible.[5] The sun should have been too cool for the planet to be anything other than a floating ball of ice. And yet there is fossil evidence of water rain. It is often said that we on Earth live in the "Goldilocks zone," where conditions are "just right" for life, since water stays liquid in this environment, but it seems that those conditions admit of many variations.

So it is far from true that the weather phenomena we experience here on Earth, even during the relatively short history of humanity, encompass the full spectrum of what weather might be.

To continue this train of thought, whatever weather is, it is obviously the case that other planets also have weather, just like Earth does – except that the similarities stop almost immediately. Jupiter stands out as the most notable example, distinctive not primarily for its size but for its giant red spot, a swirling gas storm of hydrogen and helium that has lasted at least 300 years. The famous red spot is but one of many storms on Jupiter.[6] This is a kind of weather, even though it is not something that we experience on Earth. The whole planet is a swirling cauldron of tumultuous weather-like systems, where winds routinely reach velocities of 250 miles per hour. It is the same on Saturn, another gas giant, where there are three layers of cloud, the uppermost layer of which is extremely cold – a frozen fog of ammonia ice (at a temperature of −280°F [−173°C]) – and the lower layer of which is extremely hot (close to 134°F [57°C]). The atmosphere on Saturn is mostly hydrogen, which is what gives the planet its red tinge. Both Venus and Mercury, much smaller planets still, both much closer to the sun, are blisteringly hot. Mercury is so close to the sun that surface temperatures reach 800°F (417°C) and, lacking an atmosphere, the temperatures at night reach −290°F (−180°C).[7] If Mercury's not hot enough for you, Venus's atmospheric cloud blanket of sulfuric acid boosts temperatures up to 900°F (475°C). Lead and zinc are molten at these temperatures. Earth's near neighbor, Mars, boasts considerable wind activity, causing dust storms that blast and scour the surface. On the far edge of our solar system, many of the planets are much colder and subject to what can only be described as "magical" weather. On Neptune and Uranus astronomers believe that the heavy carbon atmosphere sparks lightning storms that transform methane into coal and soot, so that it rains diamonds.[8] The temperature on Pluto drops down to −400° F (−240° C), approximately 33° Kelvin: more than cold enough to freeze nitrogen and a mere 33° away from the point at which there is no molecular motion whatsoever.

It would seem that there is even *interplanetary* weather, which is odd if you consider that there is no atmosphere in space. We can sometimes see the effects of space weather in the northern lights (aurora borealis), which are essentially coronal mass ejections (blasts of plasma and magnetic material from the sun) that impact the Earth causing the sky to light up in brilliant greens, blues, and reds. Does the aurora borealis fall in the catch-bin that we're calling weather? It certainly is an atmospheric event, but it is unusual in that it travels through space. Does it qualify as weather?

That's really just the tip of the interplanetary iceberg. There is a whole outer layer of our planet, the ionosphere, where solar radiation excites electrons, sometimes causing "ionospheric scintillation." There are in fact radiation belts around each planet. There are deviations in a planet's magnetosphere, caused by galactic cosmic rays that bombard those planets at all times. Solar winds flow outward from the sun and fluctuate much like the winds that we know on Earth. Solar radiation storms pass throughout the solar system and, though we have only a few ways of observing them, jostle each other around much like weather systems in more terrestrial contexts.

Given that weather can look so different in such vastly different environments, our search for a definition of weather also raises the possibility of different instantiations of weather occurring within our own environment. A counterfactual "Twin Earth" – a favorite device of philosophers[9] – might have us living in an atmosphere of methane, helium, hydrogen, carbon dioxide, nitrogen, XYZ, or any other of the many abundant gases in the universe. In a world of functionally similar but chemically distinct compounds, without information about chemistry or the underlying compositional makeup of those compounds, would speakers on Earth versus Twin Earth be talking about the same phenomenon?[10] Just because it isn't so, doesn't mean it couldn't be so.

With an atmosphere of a dramatically different chemical composition, weather on Earth might not just look entirely different, much as it does on other planets, but might also fall under the same description. In some future version of Earth it might also rain diamonds here, just as on Neptune, or we might bear witness to unending lightning storms and fire winds. So weather doesn't seem to be a general class of things either. Or, if it is a class, it is an incredibly large class of things, including many and most things, both actual and possible.

So then maybe instead of thinking of it as a class of phenomena, we can think of it as a kind of physical or chemical reaction, the consequences of forces of nature playing out in a universe-sized beaker. And indeed, many meteorologists study fluid dynamics and chemistry for precisely this reason: they view weather in primarily physical terms. It is therefore equally tempting to think about weather as little more than the tumult generated by a chemical reaction: as if we are all tiny particles in a giant Florence flask, riding the waves of ionic and covalent bonds.

To experience weather on this account is just to come face-to-face with a physico-chemical reaction, experiencing grand scale interactions that are, ultimately, occurring at the atomic level. Heat above us causes some elements to rise faster than others, and with them those gases swirl and lift leaves and dust, blowing through our hair, passing across our face. Cold below us causes other elements to shift into a different phase, falling out of solution and inundating our ponds.

In chemistry class we learn that chemical reactions beget "precipitates"; and precipitation is just that, the fallout of an exchange between heat and cold, the consolidation or condensation of chemical elements as they crash up against one another. As gaseous elements shift into liquid and then into solid, we experience these as the changing of seasons, as the movement and shifting of air. We feel the humid breath of summer on our necks and cheeks as vapors from the most recent rainstorm respond to the cool pavement. We experience the aching chill of winter in our fingers as we scrape the ice from our windshields. We are, in all serious respects, trapped inside a chemical snow globe.

But this physicalistic characterization of weather leaves one wanting too. Such definitions are not sufficient to capture the complexity of weather, any more than biological systems of the human body – digestive, nervous, reproductive, respiratory, musculoskeletal, etc.– are sufficient to capture the complexity of the person. To describe a person as just a migrating bag of blood and bones is to reduce personhood down to mechanical features and miss the dimensionality of a person, in much that same way that reducing weather to its mechanical features is to miss the multidimensionality of weather.

Indeed, there is much more to what weather is than just a complicated string of physical and chemical reactions. Among at least one other thing: people *experience* weather. There is something that weather *is like*. And this question about what weather is cannot be – or at least, ought not to be – limited to the best way to explain or describe weather (as if from a god's eye view), but rather about the unique and particularized *phenomenon* of weather, which is only to say that there is a phenomenological approach to this question as well.

The much-maligned branch of philosophy known as "phenomenology" seeks to explore the world not by leaning into the abstract third-person of the sciences, but to capture the world and its phenomena through the first-person lens of human experience. There's this sort of generalized question of what weather physically is, but also a much more complicated question about what rain is.[11] For most of us, though it may be interesting to learn how the weather on Jupiter or Saturn differs from the weather on Earth, it really doesn't matter when we're asking what weather is. More on that in a second.

****

At this juncture, it may make sense to think briefly not about what weather is, but about what weather is not.

A very natural way to think about weather is to think in terms of seasons. In many parts of the world there appear to be distinct timespans across a year during which differing kinds of weather occur more or less frequently. For instance, many associate winter with snow, and likewise summer with sun. April showers bring May flowers, they say, and so it is that spring is commonly associated with rain. In fact, if you look on many calendars you will see just these representations and symbols for each of the four seasons – a snowflake for winter, a raindrop for spring, a smiling sun for summer, and a leaf for autumn. Even though seasonal variation is substantial from climatological region to climatological region, people in Equatorial regions still use this convention, as if snow is coming in December.

Indeed, before the idea of climate came into vogue, people spoke of the rainy season, the dry season, the mud season, and so on. These were mostly fixed references, tied to the agriculture and the experiences of the people living, farming, and making their way in roughly one geographical location. But I think we should be careful here not to fall into this characterization of weather, as the seasons are themselves distinct from weather, and though they are tightly associated with different kinds of weather, are not weather themselves.

More contemporary parlance has us talk as if weather is synonymous with climate. But it is important (here and elsewhere) to be careful to distinguish between climate and weather as well. For most purposes, it helps to think of climate as a longitudinal phenomenon describing a circumstance in which weather events occur with greater or lesser probability. (Somebody famous – depending on who you ask, it was Mark Twain or Robert Heinlein or possibly Oxford geographer Andrew John Herbertson – once said that "climate is what we expect, weather is what we get."[12] That's not a particularly accurate definition, but it does get to the heart of the matter.) Weather, by contrast, relates to the short-term, particularized conditions of the atmosphere. Often we break climate down into different categories: tropical climates, dry climates, temperate climates, polar climates. Often we break weather down into particular events or types of precipitation: rain, snow, hail, sun.

This disconnect between weather and climate has interesting downstream consequences, not only because it is commonly the case that those who are skeptical of climate science point out the window at a recent snowfall and suggest that snow on the ground in winter serves as incontrovertible evidence that the climate is not changing. But it works in the other direction as well. We may hear the claim that climate change will result in a shift in temperature – say of two degrees Celsius – and this seems like not such a big deal, since we also experience shifts of this magnitude on a regular basis. But it is the sustained climatic shift that makes the temperature shift worrisome, not the short-term weather shift. Shifts in daily weather are not such a big deal. We know this from our own observations. We have direct experience with sometimes major shifts in weather and don't, in fact, have the experience of living in a climate altered by this degree.[13] A temperature shift in climate, however? That has downstream consequences that are much more difficult to imagine.

A Disquisition on Definition

We need at this point a somewhat more sophisticated answer to our initial question: What is weather? And the question we're asking again is not what is weather *made of*? Or what *causes* weather? Or even what *explains* weather? We want a more philosophical answer. We want to know what weather *is*.

The cheap undergraduate trick of turning to the dictionary is unlikely to be satisfactory here. *Merriam Webster's* offers a promising account, defining "weather" as "the state of the atmosphere with respect to heat or cold, wetness or dryness, calm or storm, clearness or cloudiness." The second definition of "weather" is a "state or vicissitude of life or fortune." A third definition with a slightly different sense is one of "disagreeable atmospheric conditions," such as rain, a storm, or even cold air with dampness, pointing out that we often use the term "weather" only when it gets in our way. "Weather" can also be used as a verb, of course, as in to "expose to open air." Like, we might say, "the deck furniture has been weathered." Or "weather" can also be used more metaphorically, as in "we have weathered a crisis."

Box 1 The Whether Man

One of the first people to greet the protagonist of the beloved children's book *The Phantom Tollbooth* is the Whether Man, who lives by the following creed: "I'm the Whether Man, not the Weather Man, for after all it's more important to know whether there will be weather than what the weather will be."[14]

The *Oxford English Dictionary* sheds further light on these uses, dredging up historical context to help us gain clarity on their origin. The use of the term "weather" in Old English dates at least as far back as the fifth century: "Þa het he betan þærinne micel fyr, for þon hit wæs ceald *weder*."[15] And this is interesting, at least in part, because on first glance it appears to derive from the German. In modern German, the term "weder" means neither ("weder ... noch") or either ("entweder ... oder"). I wrote a German linguist to see if there was any connection between the two uses of "weder" and he assures me that there is not, but the two are nevertheless homophones, both in English and in German. In English, the two ostensibly divergent terms "weather" and "whether" start to look a little more plausible once this connection is drawn. But what could this possibly mean? And are there any insights to glean from this tenuous connection?[16]

You are likely familiar with the ancient Greek word for time, Chronos (χρόνος), and understand it as the root of words like chronological, synchronized, chronic, chronicle, or anachronism. But you may not know that the ancient

Greeks actually had two words for time. Chronos, of course, refers to clock time, but Kairos (καιρός) refers not just to time but also to weather, as it is intimately bound up in the seasons. It relates to "timeliness," like a launch window for the space shuttle or the time when conditions are right or favorable, say, to release an arrow in archery. You wait for the wind to die down, for the target to come into view, and then you slowly release the arrow. Now is the time! The weather is right! It is time to release the arrow/launch the shuttle/harvest the crops!

More provocatively, the ancient Greek usage of Kairos is similar to its usage in modern Greek. Both languages intermingle time and weather in the word Kairos.[17] When asked how it's going, modern Greek speakers may reply with "Περνάει ο καιρός!" as in "Time keeps going," but the idea here is that this response is intimately intermingled with the idea that the clouds keep moving, that weather keeps happening. "How's the καιρός?" might be a more Anglophone phrasing, and by this may mean not only what's going on but also how is the weather? Weather, it seems, at least for Greek speakers, is ineluctably bound up in time. It is not a thing at all, but a moment, an event, and it impels us to consider what is happening in the moment.

But weather in this sense also implies something about luck and opportunity. The Greek god of opportunity, Caerus (or Kairos), is the god of luck or favorable moments. The Roman equivalent is Tempus or Occasio. Accordingly, the Greek word for "opportunistic" is καιροσκόπος, and the implication here is that the movement of the clouds, and the rising and setting of the sun, are all indications of time moving forward:

> Running swiftly, balancing on the razor's edge, bald but with a lock of hair on his forehead, he wears no clothes; if you grasp him from the front, you might be able to hold him, but once he has moved on not even Jupiter [Zeus] himself can pull him back: this is a symbol of Tempus (Opportunity) [Kairos], the brief moment in which things are possible.[18]

Indeed, the Latin word for "weather" is *tempestas*, and is likewise tied – or at least closely related to – time, which in Latin is *tempus*.[19] In French, from the Latin, *le temps*. So too in Spanish, *tiempo*. Not for nothing, such terminology has inescapable connections to the English notions of *temp*erature or even a violent storm, a *temp*est. There is certainly some dispute among etymologists about the origin of these words, but the connections between weather and time I think cannot be disentangled so neatly. These connections between weather and time persist across other unrelated languages as well. The Russian word for weather, for instance, "погода", is partly a construction involving the word "год", meaning "year" or "time". So too in Hungarian, where the word for time is "idő" and the word for weather is "időjárás."

There's a long, convoluted philological mystery to disentangle here, and I can't pretend to be even close to qualified to break it down, but I do think that these connections, even if tenuous, can be illuminating, not least because they suggest that where English speakers may not hear the connection as closely, someone speaking another language with a closer connection between the two terms actually thinks about weather differently.

So maybe it makes sense to look at more substantive discussions where such ideas might be deployed.

Weather in Mythology, Literature, and Language

Weather features prominently in other areas of ancient life as well, notably in myths and rituals.[20] Readers of this book will likely be familiar with some of the more popular gods from Western civilization. The most powerful and central of the Greek gods, Zeus, the god of weather, could strike down mortals with his lightning bolts, cause thunder to rumble across the land, and control the skies to do his bidding. His Roman counterpart, Jupiter, could do much the same. In Norse mythology, Thor, the god of thunder, wields his hammer to protect humankind with lightning, storms, strength, and fertility. But these are far from the only mythological figures with a connection to weather.

Within the Western tradition alone, there are also lesser deities associated with weather. Helios is the Greek god of the sun, tasked with driving his chariot across the sky from dawn until dusk (more on him later); Sol is the Norse god of the sun whose chariot, rather than driven by horses, is chased by wolves. There are Boreas, Eurus, Notus, and Zephyrus, who were the gods of the North, East, South, and West winds, respectively. We have of course already discussed Caerus and Occasio, the Greek and Roman gods of luck, both of whom are bound up in the seasons. Likewise, Freyr is the Norse god of fair weather and good harvest.

In the *Odyssey*, Homer describes Olympus thus:

> Olympus, where, they say, the gods' eternal mansion stands unmoved, never rocked by gale winds, never drenched by rains, nor do the drifting snows assail it, no, the clear air stretches away without a cloud, and a great radiance plays across that world where the blithe gods live all their days in bliss.[21]

When I first read this passage, I wasn't really sure what I was reading, but glance at it again. There is *no weather* (!) on Olympus. There is no rain or snow or wind, and presumably no sun, but rather just radiant light. Weather is instead a manifestation of the mortal realm, a place where gods go to play or to exact their vengeance. And in a way, this is a marvelous revelation. The gods *are* the weather: They both *control* it and *manifest as* it.

That turns out to be kind of illuminating, because weather gods are far from limited to the Western canon. Expanding out a bit, weather appears as a manifestation of the supernatural in other Indo-European mythological traditions as well. Taranis, god of the Celts, rules the skies over Gaul, Britain, and Ireland. Perun, the Slavic god of thunder, and Peperuna, his wife, the Slavic goddess of rain, for centuries dominated the weather across the Balkans. Ukko, the Finnish and Estonian sky god, controlled the weather of northern Europe. Each of these gods and traditions is different in somewhat subtle ways, but they all relate back to the force and might of the weather, which no doubt was a dominant consideration during the Bronze Age and before.

In Asia, India, and the Middle East we find related, albeit slightly different, gods. Hindu mythology has Indra (इन्द्र), king of the devas, Vedic deity of the sky, god of lightning, thunder, storms, and rain. Indra lords over all weather events much like Zeus, and like Thor, Indra wields a hammer with which he kills Vritra, the serpent of drought, and lets forth a flood of rainfall. In some later texts, Indra is not a singular god but rather a station – a title like a "king" – that changes every Hindu cosmological cycle (*manvantara*). Leigong (雷公), the Chinese god of thunder, hangs around with Yu shi (雨師), the god of rain, and Fengbo (風伯), the god of the wind. Together they wreak havoc or sow fortune across the land. The Egyptian deity Ra, with the head of a falcon and the body of a man, rules over the sun and the sky, but is flanked by more primordial gods Atum (the evening sun) and Khepri (the morning sun). These Egyptian gods of the sun – manifesting differently for different times of the day – are tasked with maintaining order in the universe. (Again we see time and weather intimately connected to one another.) Meanwhile the Egyptian deity Set is the lord of storms, an agent of chaos.

Other traditions are considerably more divergent. The dog-headed Aztec god Xolotl controls fire and lightning, but also serves to guide dead souls, whereas his friendly counterpart, Tláloc, the god of rain, brings fertility to animals and water to crops. Yet a different god in the Aztec tradition, Tezcatlipoca, lords over hurricanes, and he must have been quite the bastard – because Central America gets walloped by hurricanes every year. Chaac is the Mayan god of rain, thunder, and lightning. In Mayan lore Chaac is both one singular god but also four gods at one time, one for each of four directions from which the rain might come.[22] In the Yoruba religions of Nigeria and sub-Saharan Africa, Shango controls thunder and lightning. He is a fierce god. In almost all of these traditions, and many others unmentioned here – for example, Orko (Basque), Haikili (Polynesian), and Whaitiri (Maori) – the most powerful gods and deities were those who had control over the weather.

Set aside classical mythology for a moment and we find that more contemporary creative genres also make hefty use of weather, though more typically as a literary device than as an appeal to supernatural forces. Sometimes they do this simply, where they might use weather as an important bit of background or stage setting for a plot. Maybe they do it more actively, to move the story along. Sometimes weather is used atmospherically, to set the mood. Other times it is used abstractly, as a metaphor for emotional turmoil. At still different times it is even used as the main antagonist, as something that characters fight against. And sometimes it hearkens back to the mythos of yesteryear, turning attention to the supernatural, or to the mysteries of luck and misfortune.

Consider: "It was a dark and stormy night," begins Madeleine L'Engle in her well-known book *A Wrinkle in Time*, parodying what is commonly perceived in writing circles as paradigmatic cliché and purple prose.[23] Trite though it is, it sets the scene for the rest of the story to unfold. (This phrase was also picked up by struggling writer and intrepid beagle Snoopy [of Peanuts fame] in his efforts to transition from World War I flying ace to author.) Often this stage setting will be based in historically significant events, as with several of William Faulkner's works (*As I Lay Dying*, *The Wild Palms*, and *Go Down, Moses*), where the floods of the Mississippi Delta serve to lay the groundwork for the narrative. These works were likely influenced by real catastrophic weather events related to the flooding of the delta in 1927.[24]

So too do we find weather in poetry, as a poetic device aimed to add color and weight to a scene. The General Prologue of Chaucer's *Canterbury Tales* begins with the travels of pilgrims who are all heading to a storytelling competition:

> Whan that Aprill with his shoures soote
> The droghte of March hath perced to the roote,
> And bathed every veyne in swich licour
> Of which vertu engendred is the flour;
> Whan Zephirus eek with his sweete breeth
> Inspired hath in every holt and heeth
> The tendre croppes, and the yonge sonne
> Hath in the Ram his halve cours yronne,
> And smale fowles maken melodye,
> That slepen al the nyght with open ye
> (So priketh hem nature in hir corages),
> Thanne longen folk to goon on pilgrimages.

The April showers, the March drought, the sweet winds of spring, all serve to frame the wider story that organizes and arranges the twenty-four tales included in the collection. These same weather-related sentiments were later mirrored and turned upside down by T. S. Eliot to open *The Waste Land*, one of the most important poems of the modernist period:

> April is the cruellest month, breeding
> Lilacs out of the dead land, mixing
> Memory and desire, stirring
> Dull roots with spring rain.
> Winter kept us warm, covering
> Earth in forgetful snow, feeding
> A little life with dried tubers.
> Summer surprised us, coming over the Starnbergersee
> With a shower of rain; we stopped in the colonnade,
> And went on in sunlight, into the Hofgarten ...

Sometimes weather is deployed more fantastically, as with *All Summer in a Day*, Ray Bradbury's particularly sad science fiction story about a young child, Margot, who has moved with her family to Venus. There it is perennially cloudy and the sun appears only for two hours every seven years. Through a twist of cruelty from her classmates, she is locked in a closet and forgotten while the two hours of sun come and go. The made-for-television version was a mainstay of my childhood.

In other contexts, weather is used to move the story along. A storm may push characters out of the doldrums of their everyday life, tossing them about like flotsam on the waves. This is how Swift's Gulliver is knocked unconscious, only to find himself in the land of Lilliput. In *The Grapes of Wrath*, the Joad family is almost completely wiped out first by drought and then winter rains. They must move in response to the weather. So too in Zora Neale Hurston's

Their Eyes were Watching God, when a hurricane puts Janie and Tea Cake to their final test.

In still yet other ways, weather is used atmospherically, to color the life of the characters. In Albert Camus's *The Stranger*, Mersault endures insufferable heat on the day his mother dies, when she is buried, and when he is sentenced to murder. In the last lines of James Joyce's *The Dead*, snow falls softly outside a window as the protagonist, Gabriel Conroy, ponders his own mortality: "His soul swooned slowly, as he heard the snow falling faintly through the universe, and faintly falling, like the descent of their last end, upon all the living and the dead."

When it is not used atmospherically, it is used metaphorically, often to show inner tumult, chaos, or serenity within characters. In Shakespeare, storms illustrate the tumultuousness of *King Lear*, lightning and thunder punctuate the dark powers of the witches in *Macbeth*, and of course, a deadly hurricane lashes the ship in *The Tempest*. The thunderstorms in *Wuthering Heights* batter the farmhouse and its inhabitants as if to embody the forces outside the control of each character. In *Lord of the Flies*, the mettle of the boys is tested by a storm, wiping away the remnant bits of civilization and marking their transition into chaos and lawlessness. Tolstoy's *Anna Karenina* sits through a blizzard and struggles mightily with her feelings for Vronsky.

And of course, these uses of weather are far from limited to literature. Weather plays an outsized role in our movie-going experiences as well.[25]

Who can forget the tornado in the *Wizard of Oz*? It serves as an important plot device that provides the basis for whisking young Dorothy off to fantastical lands. Very differently, the classic Gene Kelly film *Singing in the Rain* doesn't so much use rain as a plot device but rather to set the mood, to signal the washing away of the old silent film era as the new era of "talkies" comes gushing in. In *Do the Right Thing*, Director Spike Lee, just as Camus had done to Mersault fifty years earlier, uses the incredibly hot summer of 1989 to illustrate how intense racial tensions can inflame Brooklyn's Bedford-Stuyvesant neighborhood.

Sometimes weather is used as a critical turning point for the entire assemblage. Ang Lee's film *The Ice Storm* is ostensibly built around the titular storm, though it is as much about the tumult and complexity of living through the social changes of the 1970s as it is about the weather. The award-winning Bong Joon-ho film *Parasite*, alternatively, depicts a severe rainstorm and subsequent flood that gushes through Seoul, displacing the urban poor, sending them into catastrophe shelters, while also washing away the sins of the rich.

Each additional chapter of Paul Thomas Anderson's masterpiece *Magnolia* begins with a title card describing the weather. It is only later in the film that we get a sense of why such things might be introduced, when a shower of quiz kids, misogynists, nurses, sycophants, and toadies coalesce in a thundering, if messy, crescendo. As in mythology, luck looms large across many of these more contemporary depictions of weather. (Much of that film, truly a favorite of mine, is about moral luck, about the extent to which luck affects people throughout the unfolding intersections of their lives. Given the infrequency of a string of improbable events,

Anderson makes clear that the luckiness of his subjects is not only a question about whether something has happened but also whether the people to whom it has happened can be believed when telling their stories.)

The feckless nature of the universe manifests through weather, as if to surprise on the daily. In movies like *Titanic*, *Castaway*, or *The Perfect Storm*, massive, unexpected weather events send the lives of characters catastrophically off course. Here again we see that weather both sets the stage for the story and pushes the narrative along.

Box 2 No Business like Snow Business

Classic films like *Citizen Kane* (1941) and *Holiday Inn* (1942) are reputed to have used asbestos to mimic the look of snow, unwittingly exposing their cast and crew to mesothelioma. Later films, like *It's a Wonderful Life* (1946) and *White Christmas* (1954), developed less hazardous means of creating fake snow.

Sometimes weather is even an antagonist itself, as in more fantastical films like *Twister*, *Snowmageddon*, *Icetastrophe*, *The Day After Tomorrow*, and *Sharknado*, which build on the terror that storms can induce. Such depictions of insane weather may defy meteorological science, but they nevertheless reflect the chaotic mythos surrounding weather phenomena. Or, when playing the role of antagonist, the weather functions in a more personified role, as with the horror film *Poltergeist*, in which supernatural forces swirl chaotically around the lives of mortals. There, and in many other depictions, we see weather as a force outside of the lives of others. (The "pathetic fallacy" in literature ascribes human emotions and attributes to inanimate objects, animals, or natural phenomena. Derived from the Greek word "pathos," meaning emotion, the pathetic fallacy lies in the attribution of human characteristics to nonhuman entities, as these entities are not capable of possessing human emotions or consciousness. Despite its name, the pathetic fallacy can be a powerful tool for writers, as it creates an emotional resonance between the reader and the natural world. By attributing human emotions to a storm or a mountain, for example, a writer can imbue those objects with a sense of personality and make them feel more alive.)

These are but a few of the ways in which weather appears in our creative works, but it at least gives a sense of the manifold complexity that weather provides. I'm sure you could come up with your own list. The problem is that there are so many references, so many different uses of weather, that it is virtually impossible to jot them all down.

Authors also, to be certain, use weather creatively and metaphorically, just as we all do, so much so that weather metaphors have become an invisible feature of our vernacular.

We use metaphors of weather casually in conversation to describe everything from our emotions to the state of the world around us, often without realizing that weather is in the background. We *shower* people with gifts; *blow* them away with kindness; fight *cold* wars and *hot* wars; ride the *winds* of change. If someone is unpredictable, we may describe them as *tempestuous* or *stormy*; if they are cheerful, we may say they have a *sunny* disposition; when they are sick, we may say that they are *under the weather*; when they are healthy, they are *right as rain*; when they are old, we may even be so crass as to suggest that they've *weathered*. When we face difficult challenges at work, we may feel like we're in the middle of a *shitstorm*; whereas if work is relatively easy, it's a *breeze*. When we are at the mercy of forces beyond our control, we talk about being *swept away* by events, throwing caution to the *wind*, or feeling *blown off course* by circumstances. It is particularly useful in highly volatile contexts, like war: armies *storm* beaches; soldiers duck under a *hail* of gunfire; an infantry may be *deluged*; generals get lost in the *fog* of war; citizens may be doused in yellow *rain*; relations between nations *thaw* and *freeze* over; and of course, the Nazis innovated the *blitzkrieg* (lightning war). I mentioned earlier the connections between the gods Kairos and Occasio, but metaphors of time, weather, and luck are bound up in our vernacular as well. *Lightning only strikes once*; a *windfall* is a fortunate turn of events; *the winds of fate* buffet a person around; one is lucky and skilled if one *captures lightning in a bottle*.

Box 3 Blowin' Up the Wind

The Bob Dylan song "Subterranean Homesick Blues" – in which there is a lyric exclaiming, "You don't need a weatherman to know which way the wind blows" – allegedly provided the basis for the name of the radical 1970s revolutionary group the Weather Underground (1969–1977), which purportedly aimed to overthrow the US government.

The metaphorical use of weather is not limited to personal experiences, of course. It is used in politics and other public discourse. We may enter a stormy "political climate," tossed about by volatility and uncertainty. I recently heard a political pundit suggest that "Donald Trump makes his own weather," by which she meant that whatever he does, the news media will cover it.[26] In a slightly different sense, Catholic Bishop Fulton John Sheen once said, "Each of us makes his own weather, determines the color of the skies in the emotional universe in which he inhabits."[27]

Whether we are talking about our emotions, our relationships, or the state of society, weather provides us with a rich and resonant language that helps make sense of our experiences, reminding us all at once that circumstances are often out of our control, and also subject to change. These metaphors add color and meaning to our conversations; but they also provide consolation, knowing that, like weather, our lives will eventually change.

The power of weather as a metaphor is due, no doubt, to the ubiquitous nature of weather itself, but it is more than that. It is thematically loaded. Its all-encompassing moodiness, its ambidextrous, ever-changing nature, its apparent chaos, all manifest widely across these metaphors.

Though we still as yet lack a clear definition for weather, the metaphorical connection to mood, control, and change may be important to gaining further insight into the definition we seek. Just as the weather affects our *tem*perament, it is ineluctably mood related. Just as the weather keeps us forever guessing as to how it will respond, it is also out of our control. Just as the weather can shift from sunny to stormy in a matter of minutes, so too can our emotions and the situations we find ourselves in. Metaphors of weather capture this sense of moodiness, powerlessness, and impermanence. Being buffeted by the winds of fate is a common theme in literature and pop culture, and it speaks to our deep-seated need to make sense of a world that can often seem chaotic and unpredictable. Weather is all-encompassing, chaotic, and ever-changing.

A Final Stab at a Definition

In one respect, the literary and metaphorical aspects of weather do point in the direction of a definition, inasmuch as they point to the ubiquity, power, and chaotic nature of weather. But there's still something unsatisfying about leaning too far in this direction.

We're trying to do something more here, something philosophical, and that may take quite a bit more work. *Stipulative* definitions are a common trick of philosophers that essentially skirt the dictionary problem. Rather than look to an historical, scientific, or bibliographical source for guidance, we can stipulate what we think will work for a definition, and then we can argue from there. We have to be conservative when stipulating these definitions, however, since we don't want to hoover up too many notions under our masthead. Moreover, we should be clear here that we are not so much looking for a *descriptive* definition that helps us identify weather when we see it – we'll know it when we do, as they say! – but rather a *philosophical* definition that helps us understand it. We want to know what the *essence* of weather is.

Given the preceding discussion – that weather is not a thing or a class of things, that it is not a mere physical set of interactions, but that it has phenomenological dimensions as well, that it is bound up in time – I think it may make sense to stipulate that weather is a force outside of our control.

So let's start here, with a working definition:

> Weather is a kind of *force*: a physical force on one hand, which pushes back against us, acting against us in a way that static objects do not; but a conceptual force on the other hand, which dynamically shapes and influences how we understand ourselves.

It makes sense, I think, to stipulate that weather is a force. In physics, when two objects come into contact with one another, they exert force on one another. Generally speaking, the way to model this force is to multiply the mass of those objects times the acceleration ($F = ma$). But these are just contact forces, of which there are many kinds: friction, tension, normal, air resistance, applied. There are also non-contact forces: gravitational, electrical, and magnetic. So force has this kind of mysterious association with time, space, and size. And indeed, so does weather.

There is a temporal dimension to weather, both that it occurs over time and that it changes with time. It occurs at a given time. For another, there's a spatial dimension to weather. It occurs in a given space, whether physical or real. For a third, there is a magnitude or size dimension to weather. There are large events (e.g., tornadoes) and small events (e.g., breezes).

If we consider an example of weather again, wind perhaps, and acknowledge that whatever it is that wind is, it is not just a bunch of gas molecules being pushed around but also a force that pushes against us, then that is a feature of weather that we should include in our definition. These ideas of force are borne out in literature and language as well, as we have seen the themes of mood, change, and control loom large as returning metaphors. When it is upon us, we must reckon with it.

Weather can't neatly be pointed to as a single weather event, nor can it be described more generally as precipitation. It's not tidily captured by the movement of air or on a ledger of temperatures in the shade. Rather, our best first guess at what weather might be is to think of it as a kind of external force that pushes us, and the world around us, around. In this way, it is both physical and conceptual. It is the interstitial force that makes the uniformity and consistency of our day-to-day lives far less constant.

In the absence of weather, the environments in which we find ourselves would be inert – stable states of the universe in which nothing changes. Like rocks on a cliffside, trees would sit statically, grass would stand erect, rivers would cease to flow. Like the gods on Olympus, we would know no change and could perhaps live in full control of our lives. But no. We are stuck in this world with weather. Wind comes along to stir up leaves. Rain falls from the sky to moisten grass. Lightning strikes a nearby tree. All of these events are aptly captured by weather, but it is the swirling, churning, effect-making force that soundly places them in the category.

So this is the first step in a multistep argument that I'll be rolling out over the course of this book. What I've been suggesting in this chapter is that weather isn't as easy to pin down as it may first seem. Though almost all of us are familiar with weather, and every one of us can no doubt use the term meaningfully in a sentence, when we are pushed to come up with a clear definition of weather, we often find ourselves at a loss. When we try to triangulate toward a definition, finding our way through a collection of physicalistic examples of precipitation and conceptual deployments of weather, we invariably leave a good bit unattended. Instead we can start by stipulating that weather is a powerful force that stands outside us and pushes us around.

2   Of Rain and Parades
How Weather Affects Us

The morning of September 10, 2013 began just like any other. I woke on the upper floor of my split-level house in Boulder, Colorado. Our son, six years old at the time, was sleeping peacefully in his bedroom next to ours. As I descended the stairs to brew some coffee, I thought little of the rain outside. It had been raining relentlessly for two days straight, and by this point the pitter-patter of drops on our metal roof was a barely audible, if still irritating, background static. My wife, somewhat more attuned to the world around her, followed shortly behind me. "Smells like dirt down here," she said. And I thought, "Gosh, I guess it does," so I wandered, sleepily, to the next small set of stairs to peer down into our basement.

To my great astonishment and horror, what I saw as I rounded the corner appeared first to me as an optical illusion. The walls had shrunk to half their size and the floor seemed that much higher. Our carpet had transformed to be dark and reflective, a shimmer of light winked back at me, as if a black hole had magically materialized in our basement. A smattering of partial and incomplete objects hovered peacefully across the reflective surface of this black hole. My wild hallucination lasted only a matter of seconds, which is when, with a loud gestalt "click," I realized that what was once my basement had become a swimming pool.

The 2013 floods in Boulder, Colorado swept through my home and many homes of my friends and neighbors. In our house alone it obliterated everything on our two lower levels. Our couch, our television, our guest bed, my son's toys, and numerous pillows were floating on the surface of this pool. The water in our basement had risen to the height of our door handles. A block over, our friends had water up to their ceiling. As if a reminder that life is fragile and memories ephemeral, every printed image of our prior lives – in photo albums and otherwise uncollated envelopes – was floating peacefully, along with toys and sheets and pillows and cups, in bobbing plastic bins. (Somehow my wife had the foresight only two months earlier to put all of our family pictures in those bins, where they had previously been sitting in cardboard boxes. Fortunately, those images did not get damaged, but that was a life lesson in preparedness.) Later that day, as the rain died down and the panic subsided, as we waited for pumps to become available at the local hardware store, we would eventually bust out the inner tubes and float around what was once our basement, sipping gin and lazily staring up at the ceiling wondering how the hell we were going to recover from this. (Pictures available upon request.) All told the damage to Boulder and the surrounding communities reached hundreds of millions dollars. The damage to our home alone crested $60,000.

Just as there were challenges in squaring away our definition of weather, there are challenges when addressing effects as well, specifically with regard to the way that nature affects us. So that's really what I want to talk about here. I do this in a few steps. First, it's hard to say with clarity what those impacts actually are, so I want to get some sense of the scope of both effects and impacts. Then I talk about the quantification of these impacts, about how the myriad ways in which weather affects us can be reduced – legitimately or illegitimately – to numbers. And finally, I try to think broadly about some of the more qualitative effects, turning an eye to the question of how the weather shapes us.

General Observations about Impacts and Effects

To say that weather is a force, or a force of nature, is not to say much at all. It's not *just* a force of nature. It's a force that *does something*. It has real effects. In this chapter, I advance a point much stronger than in Chapter 1: that weather is a force. I argue instead that it is a *world-creating, civilization-shaping* force, and I mean this in a pretty expansive sense.

I suspect I will get no pushback if I observe that rainwater and ice can carve mountains, that wind can prune trees or smooth rock, that lightning can cause forest fires or take out electrical grids, or that hurricanes can uproot forests and cause coastal erosion. This much is more or less accepted and given. Anybody who has visited the Grand Canyon has seen the earth-shaping might of free-running water. But it's a bit more of a stretch to argue, as I want to here, that weather infiltrates and influences everything that we do, such that weather is, perhaps more than most things around us, the single greatest civilization-shaping influence on our lives.

Such a claim might immediately leap off the page as oversimplified or hyperbolic, not least because it is not only a civilization-*shaping* force but also an *anti*-civilizing force, a force that can destroy civilizations, bring them to their knees, and leave people without places to live. I suppose it goes without saying that a force powerful enough to shape civilizations would also be powerful enough to bring them down, but I think it's precisely here that we get a sense of the peculiarity of weather. On one hand, it is foreign: an external, alien force, an agent of chaos, coming from every direction; but on the other hand, it is ever present and familiar: an intimate force without which we cannot imagine our lives.

To be clear, this is somewhat unlike other forces. Other commonly encountered forces of nature are less chaotic (more lawlike) and more familiar. Gravity is a force, for instance, that draws objects together. We kind of know how it works and we live with it harmoniously every time we put our weight on the ground. Electromagnetism is a long-range force that wows us with magic by operating along an invisible magnetic field. We may not know exactly how it works, but we have enough of an idea that we've been able to harness its mysteries to power up our computers and communicate across long distances. There are lesser-known forces too – weak nuclear forces and strong nuclear forces – that cause electrons either to stick together or to separate. These can be harnessed productively for making energy, or more destructively for blowing up cities.

There are also other categories of forces, though not of the natural type. There are military forces, for instance, which are essentially collections of individuals coordinated and directed through a hierarchy of authority that shape largely through destruction. They tear down existing buildings, infrastructure, and relationships, only to rebuild civilizations in a mold that is more reflective of those who were driving the militaries. There are also fantastical forces, such as those that confer powers upon Luke Skywalker and Yoda the Puppet. These they must channel with their minds, and though the Force was said in some retrograde prequels to be ferried through the bloodstream by midi-chlorians – itsy-bitsy, teensy-weensy intelligent life forms that reside within the cells of all living organisms (I guess) – *properly understood* the Force magically surrounds all of us and is only ever channeled through those with extensive Jedi education and training.[1] So that's at least one thing about weather as a force: more than other forces, it is in some respects powerful, chaotic, familiar, and untamable. (More on this as we go.)

Second, however, in almost all of these alternative senses of force, forces are not mere quantities. Forces are more aptly understood as vectors, which means that they have directionality: they go somewhere, they push things around. Gravity pulls down, electromagnetism attracts, nuclear forces propel in different directions, military forces march forward. The same might be said of weather: It comes and goes. It too has directionality. Its magnitude paired with this directionality contributes to its power. But the directionality and magnitude of weather matters mostly because of the way in which it acts upon us. It doesn't just do things, it doesn't just push things around, it does things *to us*, to things that *matter* to us, which is likely why it is the most natural go-to conversation starter.

*****

Consider first the nature of weather impacts: that is, what weather does to us, *how* it impacts us, and the manifold ways in which it intersects with the stuff we do.

It is, famously, a destructive force. One of the most basic senses in which weather affects us is by destroying or damaging objects that have value. It destroys homes, businesses, stores, schools, hospitals, crops, and so on. It destroys entities that have considerable intrinsic value – humans and animals and such, things that are valuable in themselves. It destroys objects that provide us with considerable utility – things that are useful to us. It destroys things that have considerable aesthetic or sentimental value – artworks, heirlooms, family photographs, and meaningful wild spaces. We don't need to get deep into value theory to see that the scope of its impact is wide. Before a storm we have something of value, and then – blink! – that thing has been washed away. This all invokes a fairly arcane metaethical discussion of value and the nature of value that I'd rather avoid here. Suffice it to say that in the policy literature, a lot of these values get clumped together as if they are comparable or similar. But further investigation reveals pretty readily that many of these values are qualitatively distinct.

For instance, its capacity to destroy *material possessions and property* is indisputable. Water intrusion will erode foundations, warp wood floors, destroy mechanical equipment (e.g., furnaces and washing machines), uproot trees, and wash away cars. As with the Boulder flood, if the water is voluminous enough it will wipe out whole neighborhoods. Freezing temperatures will crack foundations, burst pipes, push roads out of alignment, create ice dams, collapse gutters, and back up sewers. Heat waves will melt tar, ruin shingles, cause paint to bubble, and crack wood siding. Couple that with humidity and there are additional concerns about mold, mildew, condensation, and pest infestation. Wind can damage roofs, siding, fences, windows, shutters, or anything that is not pinned down. If it's strong enough it can send smaller objects flying into larger objects, downing electrical lines or accelerating fires. The devastating 2021 Marshall Fire that occurred only a few miles from my house – that I watched unfold as my neighbors' homes burned – torched more than 1,000 homes and caused over half a billion dollars in damages.[2]

Consider, next, the threats that weather poses to *human health*. Every year, the World Health Organization estimates that between 10,000 and 70,000 people die from heat waves alone.[3] Heat is by far the most deadly of the various weather-related stresses, but flooding is also responsible for considerable loss of life. During flooding events, most people die by drowning, but some are swept away, caught in landslides, electrocuted, caught in flood-associated fires, subject to carbon monoxide poisoning, and so on. There are additional spillover effects in the form of water-borne disease proliferation (with cholera, typhoid, or malaria), a problem that disproportionately affects the poor. It is hard to pull reliable numbers on these threats to life and limb, but it is at least safe to say that the populations affected more directly by weather aren't the wealthier populations. Those of us with sufficient means are fairly successful at shielding ourselves from most of the immediate effects of weather, but many in more vulnerable or more precarious positions aren't so lucky. Glancing at the news today, just as I am writing this paragraph, there have been devastating floods in Libya. On September 11, 2023, thousands were killed after two dams collapsed during heavy rains near the coastal city of Derna, adding numerous additional deaths to the toll from weather. Similar disasters happen with enough frequency that whenever you're reading this, you can go to the newspaper and read of a different deadly weather-related disaster.

Humans aren't the only ones who experience losses from weather, of course. Weather also presents hazards to *wild animals and pets*, often through drowning, overheating, starvation, or abandonment. Routinely we see imagery of families looking for their lost dogs, cats, or farm animals after a devastating storm. Invariably we know that many wild animals die from droughts and famines, whether because they themselves cannot find food or because human populations grow so hungry that they hunt wildlife to extinction. Weather also does damage to plant life, uprooting trees, breaking branches, pruning leaves, freezing fruit, and torching crops. Again, losses of nonhuman animals or plant life I think sit comfortably in a category distinct from losses of human animals, and though each are valuable for different reasons – maybe animals intrinsically and crops extrinsically – the loss of a pet or a species,

I maintain, is qualitatively distinct from its intrinsic value alone.

With loss of life and limb comes considerable *emotional and physical suffering* as well. Where some people may die of starvation due to crop loss or drought, there is often anguish and suffering *leading up to* their death. Many humans and nonhuman animals who die in weather don't just die; they die in pain. Moreover, there is unquestionably a much larger population of those who endure the pain of approaching death, or losing a limb, but manage to survive. For every death from heat, cold, wind, lightning, wildfire, drought, and debris, there are many magnitudes more injuries, resulting in immeasurable physiological pain.[4] The *experience* of pain is in this respect a very different way in which weather affects us than the loss of life or limb itself. It often *hurts* to die. And there are many people who thankfully survive but nevertheless suffer considerable agony. Add to this the distributed mental anguish, psychological trauma, family stress, and community dissolution following a devastating natural disaster from the death of loved ones, and weather's impact begins overflowing.

As if that weren't enough, along with material destruction/loss of property, loss of life and limb, and experienced pain/anguish, weather can and does destroy things that have a different kind of value to us. It often destroys our businesses, our investments, and our technologies, the loss of which deprive people the possibility of creating further value: either by earning a living or creating things that others value. It destroys *livelihoods*. Dramatic shifts in weather may cause lasting damage to factories, equipment, land fertility, or other means of production, and though these materials technically have a price tag that may be categorized as *mere* material losses, without them we may be left without a simple way to continue productivity. If topsoil erodes away in a massive flood, there is no clear market value on the soil itself, but its erosion prevents further agricultural use until the land has been restored. Extreme weather events, like drought, can shift the way that we manage our crops, forcing people who depend on crops to move off the land and find other work. Droughts cause threats to the water supply, affecting businesses that depend on the constant flow of clean water; they affect agriculture, which has downstream effects on industries that require a ready workforce; they can change the flow of rivers, negatively affecting transportation channels; they can affect the energy systems upon which many industries depend; and they can also affect overall public health. Recreation and tourism can also be affected dramatically, casting doubt on the future of certain industries. A rainy day at the beach is bad for business, just as a dry, snowless winter is bad for the ski economy. A sunny day at the beach, on the other hand, is the paradigm of a relaxing, restful, rejuvenating vacation. These dips in production then cause a cascading effect of food and water shortages, displacement, and political instability.

Weather also poses threats to *memorabilia and keepsakes*, which have a kind of sentimental or nostalgic value. When communities are uprooted by fire or flood, often what is lost is not so much the material objects themselves but more devastatingly irreplaceable items and heirlooms that can never be purchased or recovered. I mentioned this earlier, but the Boulder flood nearly destroyed my family's photographs and memories. Fortunately, my wife's knack for organization had entombed our photos in a plastic lifeboat. Given the prevalence of forest fires near our home, with some regularity we pack a "go bag." The first thing we grab, invariably, are family heirlooms and books of photographs. The digitization of photographs has changed some feelings of vulnerability, but we still value these objects over most others.

The destructive power of weather is hardly limited to physical keepsakes and sentimental losses. I briefly touched earlier on how weather is used metaphorically in art and literature. But it applies to the destructive force of weather as well, since changes in weather can wreck our *moods* or our *spirit*. Hot days leave us agitated, cold days spark cravings for soup, rainy days nudge us toward melancholy. These are real effects on us. Much as writers and artists are partly in the business of conveying aesthetic qualities through their medium, they can and do harness these shared moods precisely because the moods are real.

And finally, it is clearly the case that weather can just make an unholy mess of things. It wreaks *aesthetic* chaos as well. Beautifully manicured gardens can be shredded by hail. Majestic trees can be uprooted and downed by wind. Quaint riverbanks or seashores can be inundated and washed away by crashing waves. Well-organized downtown areas may be tossed and wrecked by tornadoes. The aesthetic values threatened by severe weather are sometimes just as important as the losses themselves, as the destructive power of nature may be more felt by looking around than just by counting losses on a spreadsheet.

I would be remiss, however, if I didn't also mention that just as weather can be a *destructive* force, it can also be a *productive* force, generating great benefits for humans and the world. Quite clearly, weather provides innumerable benefits to humanity and can impact us positively. It's incredibly important for agriculture, forestry, fishing, recreation, transportation, and construction, among other industries. But it's also true that weather provides in other ways. As we develop a new, renewable energy economy, essentially farming the weather for energy, it is equally as important, affecting our solar panels, wind turbines, hydroelectric reservoirs, and wave and tidal power stations. It supports wild plants and animals by distributing seeds, regulating growth, purifying water, and so on. It provides inspiration for artists, creating unique landscapes and natural wonders, such as geysers, waterfalls, and hot springs, and also provides fodder for us to better understand the human condition. I say more on this in Chapter 3, but the services that weather provides us – what scholars in recent years have taken to calling "ecosystem services" – often operate invisibly in the background without our acknowledgment, overshadowed, as they are, by the relentless, counterproductive push of weather.

What you'll notice about the various impacts that I outline here, more importantly, is not that they're necessarily objectively clear or comprehensive lists of all possible impacts, exhaustively citing the physical or material impacts of weather on us, but rather that I have divided the destructive power of weather up into *moral* categories, separating out different kinds of impacts that can't easily be measured across the same monotonic scale. What I mean by this is that we can't compare the pain and suffering associated with the loss of one's home against, say, the loss of a loved one. I intend to come back to these moral categories later, because I think this is ultimately one of the more interesting features of weather, and specifically of the way in which we are approaching weather nowadays. There is, most obviously, pain and suffering. That's one moral category. But there's also loss of life and limb. That's a different moral category. It's hard to compare a loss of life with any amount of suffering. Then there's livelihood destruction, which can be measured in terms of cost. This is still yet a different moral category.

Quantitative Metrics: How Much It Impacts Us

A headline screams out from the White House: "Billion-Dollar Natural Disasters Are Increasingly Common in the United States."[5] In the early 1980s, reads the report, there were only approximately two weather events causing over a billion dollars of damage per year. By the mid-2010s, this jumped to approximately 5–14 events per year. And by 2020, the number was as high as twenty billion-dollar events per year. This seems like a mighty awful number, and indeed it is, but it is more difficult to parse than it first seems.

Elsewhere a different headline shouts loudly, "US Heat Deaths by Year Hit Record High in 2023!"[6] Indeed, this is alarming as well. Record high deaths from heat? Catastrophic! Deaths from heat are indeed horrible and tell us many things about the intensity of weather. They may even tell us something about changes in weather.

Turn on the news and on any given day one can find images of hurricanes smashing boats against docks, tornadoes tearing through trailer parks, flooding in far-off lands, devastation and drought. These stories of weather battering coastlines and pushing people around play out like Greek tragedies, drawing us in as real-life dramas and windows into the horrible experiences of others. Apart from sex and fame, they are some of the most irresistible clickbait, adorning almost all mainstream news articles and attracting viewers like flies to rotting meat. Years ago, before the advent of the Internet, the weather report and the sundry, sordid enticements of video depicting destruction from storms and natural disasters would be used to keep viewers glued to the screen as talking heads droned on about the less salacious news of the day:

- "Stay tuned to see this tornado tear apart a rural school!"
- "Tonight only, we have exclusive footage of a man who saved his neighbor's mother's wife's dog from near drowning in a mudslide!"
- "Tune in at 11.00 to see Hurricane Mxyzptlk lay waste to this famous tourist destination!"

Apart from the gobsmacking power imposed by these storms, or the intrigue piqued by these headlines, a first thought you may have about these impacts of weather is that it would be good to put numbers to them. Destruction from weather events varies dramatically across time and place, so it would help to be a bit more scientific than reactionary, to evaluate this destructiveness in terms of comparable quantities.

Where in the previous section I mostly offered a rough-and-ready breakdown of relevant moral categories – damages, costs, losses, deaths, and suffering, as well as the counterpart benefits associated with weather – I didn't make much of an effort to quantify those impacts. In our highly scientized world, it seems that numbers are now the primary channel through which we make sense of, or at least compare, the relative severity of weather events.[7] So let's first look at how one might go about quantifying these impacts, focusing less on magnitudes and more on the sundry ways in such quantities are assessed.

For starters, assessments such as these often use multiple aggregate collections of information to make their approximations. In the study on billion-dollar disasters mentioned earlier, the researchers used property claim services to estimate high wind and hail damage; they used National Flood Insurance Program information to estimate flood damage; and they used US Department of Agriculture (USDA) crop insurance data to add in loss calculations (applying current market prices for various crops in order to estimate crop losses).[8] There are obvious problems with these methods, some of which the researchers note in their study, and some of which they've taken pains to correct for. Unfortunately, it's virtually impossible to make corrections for confounding variables to the satisfaction of everyone.

You may observe, for instance, that cost estimates across time spans are not simple dollars-to-dollars comparisons. Materials cost a lot less in the 1980s than they do in the 2020s, and so comparing storm damage according to cost in this way might appear to be unacceptable. Fortunately, most people doing these assessments are attuned to this concern and use instead cost-adjusted dollars, factoring in questions of inflation or interest rates that help to stabilize the comparison. No serious scholar is truly comparing storms and not cost-adjusting their assessments. What they're possibly not doing, however, is factoring in shifts in infrastructure, wealth disparities, and local responsiveness, among many other considerations, and this is where it gets mighty complicated.

A lot of other things have changed as well, making quantification slightly misleading. Medical technologies, for instance, have improved considerably over the past several decades, and so it is not unreasonable to assume that some of these health impact comparisons may be thrown off by improvements in technology. The survivability of disease outbreaks caused by weather is likely much better today than it was fifty years ago. Moreover, health disparities between different populations – the so-called social determinants of health[9] – likely play a role in quantifying losses of life and limb. Healthier, wealthier people are less likely to suffer the ill effects of weather than poorer, unhealthy people.

The same might be said of infrastructure. Technical advancements in building materials, engineering technologies, and zoning codes have ostensibly improved over the decades, and thereby create confounding complications for comparative analysis. Both health assessments and cost assessments understate damages due to such shifts in infrastructure.

This is a sort of weather-related "tree falling in the forest" puzzle: If a tornado strikes a nearby field but nothing is around to be harmed or damaged, to what extent are cost or harm metrics a reliable indicator of storm intensity? Or maybe, alternatively, a "big bad wolf" puzzle: If a storm knocks out a small village with houses made of sticks, say, how are we to make comparative sense of this storm were it positioned instead to strike a city made of bricks?

On top of these comparison problems, there are reporting and fact-finding problems. For instance, there are clear discrepancies not only in how much a storm may cost but also in the participation rates of those with insurance and how insurance data is documented. In the case of flood insurance, for instance, regional participation rates vary wildly between those who partake of the program and those who don't. Owners of single-family homes are more likely to participate in insurance markets or in studies of those markets than renters, for instance.[10]

Most complicated of all, I believe, are the confounding moral variables. Return to the categories we discussed in the previous section, to the many different ways in which the impacts of storms and weather events can be sorted into different bins. Again, we have multiple moral categories that, on initial assessment, can be clumped together as losses, but upon further analysis make us uneasy.

Comparing lost lives with billions of dollars in damage is, on many accounts, an apples to oranges comparison; or maybe even an apples to orangutans comparison. It's not just that the values are *incomparable* but that they are in fact *incommensurable*, meaning that they don't overlap in a conceptually defensible way. This is the reason that I introduced the various moral categories in the previous section. Damage assessments from weather events are controversial not only because they are often flimsy proxies for that which they aim to measure, but also because it is not clear that they offer the appropriate *moral* comparison.

It is tempting in such circumstances to respond with the observation that, even if illicit, we do as a matter of practical reasoning make such value comparisons all the time – every time we get in our car, for instance, we have effectively made a decision about the trade-offs of safety features versus the value of our life and the lives of our family members. At the same time, there's a more basic sense in which, though it may be true that we do make these kinds of trade-offs all the time, there's something perverse about framing our decisions in such a way that these categorically different values are factored out.

More distressingly, these kinds of impact comparisons can be and are regularly obfuscated by strategic actors – politicians, industry lobbyists, insurance companies, etc. – who inflate or deflate impact reports in order to take advantage of these tensions. It is one thing to use a metric like temperature to measure some aspect of the weather, which is tied not so much to the impacts of a weather event but rather to the weather event itself, but another thing entirely to use a metric like cost or health or loss of life. There is considerable research suggesting precisely this: Quantitative metrics that purport objectivity, and indeed masquerade as impartial measures, are particularly vulnerable to political exploitation and doubt-casting.

In the highly politicized climate debates, comparative reports of the sort mentioned here, and their related objections, are often deployed to reinforce a position related to the trustworthiness of the climate science or to undermine it. I'm not going to enter into that discussion here. I want to be careful to avoid any implications one way or the other on climate. That's not my focus in this book, and we saw earlier that we should in many cases evaluate weather independently of climate. I do, however, think we have to ask serious questions about the aptness of the comparisons. What is the signal-to-noise ratio on these metrics? How reflective are these metrics of the underlying reality of impacts? More importantly, what is the political cost in employing them *as if* they were objective comparisons between storms?

Given the problems with quantification and, ultimately, the vulnerability of these metrics to politicization, what do numbers such as these tell us about the impacts of weather? How are we to disentangle, for instance, problems with human built infrastructure, or lack of social support networks, from the intensity of the weather itself? When we rely on quantitative metrics such as these to help illuminate or communicate important impact information, we open a

Pandora's box of doubts that, whether supported or not, undermine the effort to make sense of these claims.

*****

In the interest of finding a more stable and acceptable metric, perhaps the best place to start is with direct, unmediated meteorological measurements themselves: wind speed, wind direction, rainfall, temperature, barometric pressure, wind chill, relative humidity, heat index, cloud cover, air quality, cloud variety, altitude of systems, pressure trends, and so on. All of these taken together offer a more direct and seemingly objective measure of some feature or other – strength, size, intensity, duration – of a weather system. At a minimum, they offer at least some sense of how weather events might affect our instrumentation. On very hot days, the mercury will be excited and will expand. On very cold days, the mercury will contract. If we're clever enough about it, we can put a bunch of hatch marks on our glass tubing and record a temperature.

From there then we can come up with different scales and systems to explain the intensity or severity of storms. And there are actually quite a few such scales. The Saffir–Simpson Hurricane Wind Scale (Table 1) rates hurricanes on a scale from 1 to 5, with a Category 5 hurricane being the strongest. The Beaufort scale instead uses observational data to relate wind speed to existing (mostly ocean) conditions, creating a hierarchy of wind types from calm to breeze to gale to hurricane, each associated with ripples, wavelets, small waves, large waves, or foam and spray. The

TABLE 1 *The Saffir–Simpson Hurricane Severity Scale.*

Classification	mph	km/h	Impacts
Category 5	≥157	≥252	**Catastrophic:** heavy damage to buildings; extensive wind damage; framed homes destroyed.
Category 4	130–156	209–251	**Extreme:** major structural damage and heavy flooding; evacuation required; power outages for weeks.
Category 3	111–129	178–208	**Devastating:** structural damage to small houses; many trees uprooted; inland flooding.
Category 2	96–110	154–177	**Extensive:** roof and siding damage to buildings; shallow trees uprooted; power outages.
Category 1	74–95	119–153	**Moderate:** limited structural damage; large branches down; damage to power lines; some flooding.
Tropical storm	39–73	63–118	–
Tropical depression	≤38	≤62	–

Enhanced Fujita Scale offers a similar rating for tornadoes. There's the Dartmouth Flood Observatory Flood Magnitude Scale that looks at flood height, water velocity, and water discharge to approximate the impacts from flooding.[11] There are synoptic scales to measure atmospheric pressure.

All of these scales are put together using measurements that themselves are essentially meteorological, inasmuch as they ostensibly tell us only about the storm itself. They aren't as clearly or directly connected to the impacts of the storm. We are left instead to extrapolate from the metrics how the weather will affect the human population. But what you notice upon further examination is that in many cases even these ostensibly objective scales are designed in many ways to tell us how weather will affect us. At minimum, though each measurement is divvied up into a more discrete category, the categories themselves reflect how the storm impacts some aspect of the world. And this much even plays out in efforts to clarify.

Though the scales themselves relate mostly to wind speed alone, when these scales are communicated to the public, it is almost always is the case that they are translated into more approachable "human impact" terms. For instance, the National Hurricane Center describes the wind scale in terms of the damage that will eventuate. For Category 5 Hurricanes with a wind speed of 157 mph or higher:

> Catastrophic damage will occur: A high percentage of framed homes will be destroyed, with total roof failure and wall collapse. Fallen trees and power poles will isolate residential areas. Power outages will last for weeks to possibly months. Most of the area will be uninhabitable for weeks or months.[12]

They offer these translations for each of the different categories of hurricane.

So it seems that even with relatively direct measurements of the various aspects of weather, we must return in some respect to ourselves. Says Protagoras, "Man is the measure of all things," and by this he means not just that we are the ultimate arbiters of what is being measured but that all measurements can be traced back to our perceptual apparatus and the relevance of the things we are measuring to us and our lives. This is not to say that there are not real atmospheric effects of weather to measure, only that often the metrics that we use to measure those effects are tied tightly to the things we care about.

Which points to another area in which our thinking about weather has changed: the study of weather *communication*. Meteorologists and reporters have discovered that it really matters how they communicate impending weather events. Depending on what they say, this affects what other people do. When they use terms such as "megadrought," "bomb cyclone," and "storm of the century," partly directed at garnering eyeballs, they lean toward the hyperbolic, amplifying the intensity of storms simply by labeling them as something new and exciting. While there is consensus among many in the scientific community that some weather events are getting more frequent and more intense, partly due to climate change, the labeling of these weather events raises concerns that the science itself is hard to parse given the rampant proclivity to hyperbole.[13]

This is true not just about what is said and how it is said but also about the medium in which it is communicated. Where the nightly news fifty years ago would make use of flashy weather reports to try to keep viewers tuned in, this eventually morphed with the so-called 24-hour news cycle and became repeated chyrons and warnings beneath regularly scheduled programming. Social media expanded our capacity to send out weather forecasts even further. Weather reports are now virtually everywhere: on website banners, on Twitter, next to ads on our Facebook feeds, etc.

These two factors – medium and message – affect our perceptions of weather and ultimately feed into how we think and talk about the risk that weather poses to us. So there are the impacts of weather, and there are the perceived impacts of weather, and the two of these may or may not line up.

Qualitative Effects: How It Shapes Us

As before, it is important to see how broad-reaching these impacts are, not just in terms of how we can assign magnitudes and quantities to them, but also regarding their indirect, downstream shaping effects.

In a very trivial sense, I observed earlier how, when weather shapes our environment, it also by turn shapes our industry. Over time, rivers will shape canyons, ice will break rocks, and water and rain will affect plant life, changing the ecology. Depending on the available resources and ecosystem services – sunlight, rain, snow, wind, temperature – industries will coalesce around these resources. Agriculture may find a happy home in open plains, where weather can roll across the grasslands and gently kiss the crops. Tourism may plant roots in the mountains, where visiting daytrippers go for hikes or spend time skiing. Cities may spring up on the banks of rivers that provide commercial transport options both inland and to the sea. This obviously has an effect on who moves where, what jobs are available, whether a region becomes more urban, rural, or suburban, and so on.

There are entire scholarly journals dedicated essentially to the effects of weather on important events in history, laying out with specificity the extent to which environmental factors have contributed to the current shape of things. The journal *Environment and History* is overflowing with articles on various storms or droughts that influenced or shifted the historical course of a given region.[14] Napoleon's army allegedly met its fate at Waterloo largely because of the weather, when the rains caused the battlefield to turn into a mud bog.[15] But even leading up to that, during Napoleon's Russia campaign, the summers were blazing hot, draining soldiers, while the winters were frigid and deadly.[16] So too with the German army marching across eastern Europe toward Russia during World War II.[17] The timing of the Normandy invasion was almost entirely dictated by the weather forecast, as allied forces anticipated favorable weather conditions; the German forces, relying on a different weather forecast, were taken by surprise. The Japanese attack on Pearl Harbor is also said to have benefited the Japanese since there was allegedly good visibility with only partially cloudy skies – perfect for an air assault.[18] On the flip side, the atomic bomb that the US dropped on Nagasaki, Japan, was originally scheduled to be dropped on a different city, Kokura, but that target was scuttled because of – you guessed it – cloud cover.

The list of weather-related military redirections hardly stops (or even starts) there: the Mongol invasions of Japan were said to be thwarted by a major storm that threw

Kublai Khan's ships off course;[19] the terrible grain harvest from the hot summers of 1788 is thought to have led to the French Revolution in 1789;[20] the frigid winter of 1657 to 1658 allegedly affected the Swedish army crossing over into Denmark;[21] the famines in Finland and Estonia in the late seventeenth century left many with no choice but to turn to cannibalism;[22] the Crimean war in the 1850s was seriously affected by fog; the mildness of the winter during 1672–1673 is thought to have helped the Dutch prevent the French conquest of Amsterdam.[23]

Those are some of the more dramatic cases where weather has shaped history, and therefore the landscapes of cities and industries. But there are considerably more mundane examples that don't involve leveling cities to make way for new political regimes.

My hometown of Boulder, Colorado is a great place to live – with a thriving university, a booming tech economy, and a bustling downtown area with restaurants and shops – but we have benefited tremendously from our proximity to the mountains. People come here at least in part for our recreational offerings, which in most cases means that they come here to ski in the winter and hike in the summer. If Boulder were in a different part of the country with different weather, we would likely have a very different student body at the university, perhaps one inclined to sunbathing and swimming. Winter snows play an enormous role in this. If the snows were to disappear tomorrow – say, due to climate change, or perhaps due to the intervention of Heat Miser, the snow-loathing supervillain from the 1974 Christmas special *The Year Without a Santa Claus*[24] – it is more than likely that our attraction as a hub of ski activity would also go away. That's not to say that the *town* would cease to exist, or that the Front Range of Colorado wouldn't continue to be a hub of economic activity. It just might be a *qualitatively different* place, one more focused on warm weather mountain-based activity. Suffice it to say, weather shapes our social environment.

In many cases the shaping effects are considerably more subtle, only forcing a slight nuance in our thinking about where we are and who we are. It never ceases to amaze me, for instance, how green Colorado can get come spring, when the rains have woken the grass that lies just underneath the brown, dry turf. There are days when it feels to me that I am living in a different place, a world long forgotten after our windswept winters. Sun alone, passing through clouds or bearing down harshly from above, changes how we experience and see the world around us, and insofar as it changes how we see the world around us it changes how we see ourselves within that world.

Among the many hats I wear, I also have a side-hustle as a professional freelance photographer. In that capacity, I pay close attention to weather, location, and light when deciding when to pull out my camera. If it's sunny, cloudy, overcast, foggy, sunset, sunrise, blue hour, golden hour – each of these factors influences how my camera sees the world, how I take it in, and ultimately how the pictures eventually look. I do this in part because photography is in its most basic sense the project of capturing the world and giving an impression, of creating images that reflect the mood or idea. Seeing has long been a metaphor for knowing, and indeed, if the weather plays such an important role in the way we see the world, as I attest that it does, then it also plays an important role in the way in which we understand or know the world.

In turn, this light, and light associated with different kinds of weather, can have substantial impacts on our mental health. Long, dark winters are thought by many to contribute to seasonal affective disorder, leaving people with a lack of interest in many activities, social withdrawal, sleep problems, irritability, anxiety, and just general feelings of hopelessness. Likewise, warm sunny summers are said to lift our spirits. This much has been understood for centuries, and over the years, people have sought to counterbalance the psychological effects of weather by taking steps to perk up the afflicted. The German Catholic Priest Sebastian Kneipp (1821–1897), one of the earliest naturopaths, pioneered a method of therapy known as the "Natural Cure Movement" in the nineteenth century that aimed to take advantage of the natural healing powers of sunlight, air, water, humidity, pressure, and other features of weather. Quackery though this may have been, it resulted in a lasting legacy that enables German doctors to prescribe state-subsidized medical "spa breaks" at many of the resorts around the country.[25] The bit that is not so much quackery, however, was that the German winters and weather were known to bring people down, and the spa break, even if it didn't cure a person of more biological ailments, could at least provide welcome respite to those who were struggling to get through the winter.

The scientific findings on such an approach are decidedly mixed. A 1984 study found that weather can affect a range of different mood variables, including "concentration, cooperation, anxiety, potency, aggression, depression, sleepiness, skepticism, control, and optimism."[26] A 2005 study suggested that "pleasant weather improves mood and broadens cognition in the spring because people have been deprived of such weather during the winter."[27] Even more recent studies have suggested that weather, when it does affect our personality, affects us primarily in the negative direction, not so much in the positive direction.[28] Nevertheless, one of the other medically accepted treatments for seasonal affective disorder is through artificial light therapy, not just increasing exposure to external sunlight but also by using light bulbs that emit warm, daylight temperatures.

More recently, researchers have explained these mood-altering features of weather in medicalized terms, suggesting that the reason the sun does this to us is that it stimulates the body's vitamin D production. And so a whole vitamin industry has developed around the idea that by supplementing vitamins we might otherwise produce naturally we can help people break out of depression, giving them greater mental focus.

So it seems that weather shapes our political environment (by redirecting our armies), it impacts our social environment (by affecting our industries), it colors our perception of the environment (by shifting how the world is reflected back to us), it affects our psychology (by altering our moods), but it also affects who we are.

And that's where it gets quite a bit more complicated.

Environmental Determinism

Sometimes, when temperatures drop below freezing in Colorado, I watch the birds, squirrels, and deer, and I wonder how they can survive in these conditions. It seems impossible. But clearly, animals have a range of adaptations that make it possible for them to survive harsh weather conditions. Bears, we know, hibernate, as do numerous other mammals, like hedgehogs, skunks, some bats, ground squirrels, marmots, and fat-tailed dwarf lemurs. Hibernation isn't limited to mammals. Wood frogs, bumblebees, snails, box turtles, and some snakes hibernate too. There's actually a bird version of hibernation called torpor – not technically hibernation – that a single bird, the common poorwill, is known to undergo. Still others change the color of their bodies to camouflage themselves as snow sets in. Snowshoe hares, weasels, arctic foxes, and ptarmigans all do this. Other animals develop thick fur or take on a bunch of fat to protect themselves for the winter. The same goes for hot weather. Some animals, like toads and snails, enter a hot weather form of hibernation called estivation. Animals have developed techniques such as panting, sweating, shedding, showering themselves with water, and burrowing underground in order to beat the heat.

It makes sense that the same might be true for us. Sunnier weather can affect our skin, giving us sunburn, sun tans, or cancer. Over time it might make us and our offspring slightly darker. Colder air makes our joints feel stiff whereas warmer air can help us feel less pained and more likely to get outside and move. Rainy weather may cause inflammation or lethargy, causing us to be less industrious. Over time, we might develop resistant traits that respond to these effects. A lot of weather has to do with water, and water makes up about 72 percent of our total body weight. Our bodies rely on water to regulate temperature, and so when the temperature changes, our bodies change in response. Accordingly, we change our activities, working out more in the summer, becoming more sedentary in the winter. These activities in turn affect our mental health, our physiology, our outdoor activities, and so on.

Those are just some of the most obvious direct and immediate ways in which weather shapes who we are and how we live, and in many respects they're uncontroversial. But there are much more broad-reaching and contested theories that purport not just to understand the qualitative ways in which weather shapes how we live and what we do but also to understand how weather, and nature more generally, shapes ethnicity, culture, race, and even national identity. Such theories fall under the general masthead of "environmental determinism," and at least by my estimation make stronger claims than simply that weather has qualitative shaping effects on how we live our lives.

The idea that the environment influences or determines who we are is nothing new, of course. It's been around for thousands of years, taking various contours as beliefs about the person have changed. Determinism proposes exactly what it sounds like: that we are *determined* by our environment; that the places we live in, the weather we experience, the vapors we breathe, all factor into our ultimate build. Acknowledging what we said earlier, it can take on a physiological, physical, mental, or even cultural dimension. Where it is one thing to note that the weather may impact and constrain our lives in some way, affecting what we see or how we feel, the determinist wants to hold more strongly that what we see and how we feel translates over into regionally or culturally specific practices. So, for instance, where almost all of us can acknowledge that hot sunny weather makes us less inclined to run around and be active, perhaps inspiring us to seek shade and shelter, the environmental determinist will try to explain culturally embedded practices in terms of these weather phenomena. They may say, for instance, that people who are raised and live in hot, humid environments are more inclined to laze around during the middle of the afternoon, perhaps not seeking work, whereas those who are raised and live in cold frigid environments, with long difficult winters, are more inclined to be industrious, to prepare for the winter, because they know that their life depends upon it. This kind of thinking has resonance in the observations I made about the adaptations of animals. That is, the idea is not just that the weather *affects* who we are, but rather that we are either born naturally to endure specific environmental conditions or we adapt to live with them.

It is important to see, however, that these versions of environmental determinism are nuanced, and that they take various shades of plausibility depending on the extent to which they overreach and bridge into essentialism. At least one thought, common in the works of Hippocrates (460–377 BC), Herodotus, and Avicenna most notably, but others as well, including Dicaearchus and Agartharchides, was that the weather determines the health of those who are the subjects of it. Such a notion is not unfamiliar to any parent who has insisted that their child bundle up before walking out into the cold. This environmental determinism appears, among other places, in Hippocrates' book *On Airs, Waters, Places*, where he argues that the health of the person is influenced by air, water, and places:[29]

> 1. Whoever wishes to investigate medicine properly should proceed by first considering the seasons of the year and what effects each of them produces, for they are not all alike, but differ much among themselves as regards their influence. Next, one should study the winds, the heat and cold, especially values which are common to all countries, and then those which are peculiar to each locality. Similarly, when someone arrives in a city to which he is a stranger, he ought to consider its situation as regards the prevailing winds and the rising of the Sun; for its influence is not the same if it faces north or south, or if it faces the rising or the setting Sun (90).
>
> 24. And there are in Europe other tribes ... such as inhabit a country which is mountainous, rugged, elevated, and well watered, and where the changes of the seasons are very great, are likely to have great variety of shapes among them, and to be naturally of an enterprising and warlike disposition; and such persons are apt to have no little of the savage and ferocious in their nature; but such as dwell in places which are low-lying, abounding in meadows and ill ventilated, and who have a larger proportion of hot than of cold winds, and who make use of warm waters- these are not likely to be of large stature nor well proportioned, but are of a broad

> make, fleshy, and have black hair; and they are rather of a dark than of a light complexion, and are less likely to be phlegmatic than bilious; courage and laborious enterprise are not naturally in them, but may be engendered in them by means of their institutions. And if there be rivers in the country which carry off the stagnant and rain water from it, these may be wholesome and clear; but if there be no rivers, but the inhabitants drink the waters of fountains, and such as are stagnant and marshy, they must necessarily have prominent bellies and enlarged spleens (220–221).[30]

These deterministic ideas were further built out and developed by the Roman physician Galen of Pergamum (129–216 AD), who subscribed to a view of medicine known as *humorism*, in which the elements of the body could be divided into the four humors – phlegm, blood, yellow bile, and black bile – all of which should be kept in balance. Should these humors – essentially fluids – fall out of balance, they would accordingly create different temperaments in a person, leaving them phlegmatic, sanguine, choleric, or melancholic. As a consequence, Galen was also a big advocate of bloodletting – basically, rebalancing the humors – which he believed not only had to be done according to one's illness but also had to be attuned to the weather and temperature outside.

Likewise, from later Islamic scholarship, we find similar deterministic ideas in Avicenna's *al-Qanun fi al-Tibb* (The Canon of Medicine), in which Avicenna argues that:

> Healthy air remains clear unless there be admixed with it vapours from lakes or from stagnant or deep waters or

> marshy lands, or from places where potherbs are cultivated – especially cabbages and herb rocket; or where certain resinous trees or trees of bad temperament (box, yew) grow, or where nuts or figs grow, or where there are offensive odours and evil-smelling winds ... such air is not retained deep in the earth. It becomes warm quickly after sunrise, and becomes cold quickly after the sun has set ... Air is good when it does not interfere with one's breathing or cause the throat to contract.[31]

It beggars the imagination how one might live among cabbages and hope ever to breathe freely!

A more insidious version of environmental determinism lends itself to support nationalism, in which the very soil and the air of a nation, the *terroir*, makes people or the ethnicity of that nation healthy or good or strong – or in some way "better than." The theoretical precursors of these views have roots to some extent in political theory. Hesiod developed the notion of the metallic races, in which he posits a Golden Age, when men were immortal, living among the gods on Olympus.[32] Plato too has his own thoughts in the Myth of the Metals (one aspect of the famous Noble Lie), suggesting that in order to maintain social order, the population should be told that they are all made of the same earth, that people of more noble lineage have higher concentrations of gold and silver than people of less noble lineage, who have bronze in their blood.[33] In these instances, the gods determine through their choices the various different races and their strengths and weaknesses. Whole nations are given bread and honey, while others are left to starve.

Many of these politically motived environmental determinist ideas fell out of favor during the dark and Middle Ages, as theorists held mostly "static theories" of human attributes and cultures. That is, they held that the features of a person were pre-formed at birth, presumably by God or the gods. The idea was simply that an almighty god would have control over us as individuals and us as peoples, and any implication to the contrary would in some way suggest that God was not in control of our destiny.

At least one exception might be found in the work of the Jewish scholar Rabbi Moses Ben Maimon (1138–1204) – aka Maimonides, aka "the Rambam" (רמב״ם) – who suggested that medical ailments and weaknesses, such as asthma, were closely associated with climate and diet. This is, obviously, not a particularly controversial observation alone, as a lot of it still holds true today, but his view was that we could seize control of our medical future by changing the environment in which we were situated. Though not a full-throated determinism, inasmuch as Maimonides seems to encourage people to educate themselves and take control of their health, this suggests that environmental factors were at least partly determinative of health and vitality. What is curious is that sometimes Maimonides is suggested to have advanced some of the more racialized conceptions of environmental influence.

Fast forward to the Enlightenment where, among others, folks like Nicholas Malebranche (1638–1715) and Georges-Louis Leclerc, Comte de Buffon (1707–1788) carried forward these static theories of human attributes and argued that the static human features were produced, not only by God, but instead by climatic or environmental differences.[34] Charles-Louis de Secondat (1689–1755), aka the Baron de Montesquieu (or just Montesquieu), argued that meteorological conditions had an effect on the bile of humans, resulting in more and less successful moral and political systems.[35] Even the philosopher David Hume, on one hand skeptical of the idea that there might be climate-based racial differences that could account for differences in temperament, nevertheless did maintain, in his essay "Of National Characters," crass views characteristic of the superiority of white races (what has been termed by some "philosophical racism"):[36]

> The only observation, with regard to the difference of men in different climates, on which we can rest any weight, is the vulgar one, that people in the northern regions have a greater inclination to strong liquors, and those in the southern to love and women. One can assign a very probable *physical* cause for this difference. Wine and distilled waters warm the frozen blood in the colder climates, and fortify men against the injuries of the weather. As the genial heat of the sun, in the countries exposed to his beams, inflames the blood, and exalts the passion between the sexes.[37]

All of this may sound fairly uncontroversial and innocent, and indeed much of it was a product of the time, but the important bit is that determinism, when fully embraced, strips agency from the community of actors, removing considerations of self-determination that otherwise add justificatory color to history. To suggest that people are just pushed along by the world around them, as if they were pawns in a game, moved either by nature or the gods, suggest that they don't have any control over what befalls them. In some of the social sciences literature, the distinction is generally between determinism and "possibilism," which is the idea that although humans are never entirely free from the influences of their environment, there's at least some space to allow for agency.

When paired with some of the later, more "scientific" assumptions of Lamarckian and/or Darwinian natural selection, environmental determinism takes on an even less intentional, and more accidental, dimension. Here it is not the gods who determine the composition of people but rather nature itself (in the form of genetic selection) and the extent to which whole, adapted, and fit races emerge out of particular geographical locations. Some of these ideas were used to justify the dominance of empires, the incursion of colonialism, the deliberate selection of eugenics, and eventually the forced extermination of Nazism. Adolf Hitler – hardly a thinker worth repeating but at least a popularizer of ideas – offers perhaps the starkest example of a means through which environmental determinism, sanitized through science, can be used for insidious purposes. The *Blut und Boden* (blood and soil) position of the Nazis and their related notion of *Lebensraum* (living space), articulated earlier by Friedrich Ratzel (1844–1904), effectively suggested that the German people, made superior in part by their environment, could expand their racial dominance by grabbing more territory and land to grow the Aryan nation.[38]

That's all pretty extreme, and fortunately most of these ideas have fallen out of favor (at least in intellectual circles). Nevertheless, some ideas persist in considerably softened form. Jared Diamond's widely read *Guns, Germs, and Steel* advances a soft or neo-environmental determinism, suggesting not so much that people are inherently or genetically composed in such a way that has them adapted to their environment, but that certain economic and political institutions can be attributed in part to environmental factors.[39] This is far less controversial – but controversial nevertheless.

So let's return to our working definition and add a bit more flesh to these bones:

> Weather is a *destructive and productive* force: a force on one hand that obliterates value in the world, natural and artificial; but a force on the other hand that creates value as well, providing the basic building blocks for life and shaping people and civilizations.

In this respect, weather is, as I mentioned at the beginning, a civilization-shaping force that imposes itself upon us at all times, that surrounds us and shapes us, that turns us to look for methods of minimizing the unwanted impacts and taking advantage of the more desirable impacts.

The point of this chapter's exploration has been that the natural forces that act around us also act upon us, affecting us in ways that otherwise go unnoticed. In the post-Enlightenment context in which we currently live, we are all the more inclined to make sense of these impacts in terms of discrete, differentiated metrics. But it is also easy to see that these metrics only capture a small portion of a much bigger picture. They leave out many of the human dimensions of experiencing and living with, or being impacted by, weather. The "ineffable remainder" of its qualitative shaping effects stays masked behind the appearance of objectivity. They collapse the irreducible moral categories that I began this chapter with, suggesting that one of the ways to make sense of destruction is in terms of cost or loss.

One objection that may have been nagging at you is that there is an as-yet unjustified moral valence to these destructive and productive aspects of weather; that to talk in terms of destruction or production is to smuggle in value positions – ideas of good or bad – that aren't always so clear. Destructive forces are bad. Productive forces are good. Doesn't this really depend a lot on our perspective? This is what I deal with next.

3 Battening the Hatches
Living with Weather

One rainy day years ago, when my son was in third grade, we prepared to walk to the bus stop. As one wont to do on rainy days, I reached into our closet and pulled out an umbrella, an unremarkable object if ever there was one.

Now then, you may be getting the wrong impression given the story I last told about the torrential rains that would inundate our house in the Boulder floods of 2013. Colorado is an arid environment, and though it rains here, it generally doesn't rain much or for long. Needless to say, as my son was born and raised in Colorado, he was relatively unaccustomed to rain. It hadn't really occurred to me how unaccustomed he was, however, until I brought out this umbrella.

Umbrella technology – this boring, unassuming technology – that had otherwise been sitting in my closet, quietly waiting for rain, suddenly emerged as a marvel of civilization. I brought it out, put it on the stoop, and opened it up. My son's eyes grew wide. He was fascinated. Never had he seen such a cool contraption. He wanted to see me open it and close it and open it again. We walked together to the bus stop, under the umbrella, and not only was the umbrella unfamiliar to him, but it was also unfamiliar to many other kids around the bus stop.

Naturally I and the other parents got a kick out of this. It's not like we live in a desert. Though Colorado is technically located in the "high desert," it's not as if it never rains here. It's just that usually when it does, it lasts for a short time, so we can wait it out, or it evaporates rapidly. It's hard to blame the kids. We've all heard stories of people living in southern climates who have never experienced snow, or people who live in the north and are so pasty-faced and unfamiliar with the power of the sun that they walk onto the beach and turn lobster red. The idea that the people around us may never have seen technology of the sort that protects us from rain was hysterical. Umbrellas are so ubiquitous, so prevalent, that the thought had never crossed our minds.

In some sense this is true about all technology, that when widely adopted and incredibly useful, it just fades invisibly into the background. Artificial light, air conditioning, indoor plumbing, electrification, refrigeration, gas stoves, water filtration, ambient Muzak. It takes a moment of reflection to acknowledge or recognize the extent to which these technologies are doing a lot of work for us. But there's a second important sense in which many of the invisible technologies that permeate our lives were developed to respond to the weather. The destructive and productive capacities of weather, what weather takes away and brings to our lives, is partly determined by how well equipped we are to manage it. It is true that weather shapes how we live our lives, but whether we accept whatever weather throws our way or develop technologies to protect ourselves from or harness weather also determines in part how we understand

and make sense of it. And our technologies are themselves born of perceived necessities. Without a need, without a reason to develop these technologies, the thought simply never crosses our mind that we might need a device to do this for us.

Reframing weather as something other than a mere force, in terms of both its productive and destructive nature, leaves us with yet a new batch of questions, largely about how we might respond to this force. In Chapter 2, I focused mostly on impacts, on how our lives are qualitatively and quantitively affected by the weather. That may seem like much of what I aim to talk about here. But those were mere shaping effects. In the next several chapters, I look at the ways in which we have altered our relationship to weather through innovations in technology and knowledge. In this chapter, I explore the ways in which we've tried to live with or respond to the weather. Obviously the central issues here are not only the extent to which we have come around to and learned to live with weather but also the technologies that we have developed to beat back the harms of weather, or to harness the power of weather.

Dispositional Orientations toward the Weather

To advance this position, I discuss what may be best described as "dispositional orientations" toward the weather.[1] Specifically, I look at the way in which our various technologies and innovations feed back into our attitudes about – that is, our dispositional orientations toward – weather. So far as

I can tell, there are at least three possibilities: we can accept it, we can resist it, or we can harness it.

The first of these is somewhat more passive than the other two. What it would mean to accept weather is just to allow it to wash over us, to live in the world essentially riding the peaks and valleys of weather's destructive and productive might. You might think of the one kid on the playground who continues to play, unbothered by the pouring rain. The last two are effectively two sides of the same coin: active responses to the destructive and productive power of weather. Here you might think of the kid who runs for shelter, or the other kid who relishes the rain: collecting it, enjoying it, playing with it. Taken together, however, I think these three attitudes tell us a story about what weather is, and perhaps more specifically what weather is *to us*.

We see evidence of each of these attitudes at all times over the course of history – from the very beginning to the present day – but what I think is interesting is how our technologies and ideas have slowly shifted these dispositional orientations, thereby, in effect, shifting the moral valence of weather: from an abiding and serious concern to a much more innocuous background phenomenon to maybe even one that we see as beneficial. At the same time, the transition has not been clean or direct. Our relationship with weather has not been a linear transition from nasty to nice, but rather one in which the different aspects of weather vacillate between both nasty and nice, with a whole bunch of intermediate phases that leave our relationship fraught.

For centuries, weather was a real, nonnegotiable aspect of survival: the kind of thing we couldn't ignore.

It posed serious threats to our well-being, both in terms of what it could deliver and in terms of what it might fail to deliver. But it also provided for us. Slowly, with development and eventually industrialization, it has morphed into the kind of thing that we don't think much about – until we actually have to think about it, which is when it reminds us of its force. What is tricky here is bearing in mind that each of these orientations does not so much carve out a distinct phase of human development, but rather points to attitudes that we all take at different times depending on the circumstances – maybe even within the same weather event. We may embrace the rain but resent the inconvenience. The central problem for us, then, is to try to understand how our dispositions and attitudes might have shifted in the face of changing historical and cultural contexts.

Accepting It

There is a certain respect in which the weather surrounds us and just kind of – is. It does its thing, we do ours, and since we have so little control over the weather, so little sense of where it will go and what it will do to us, the best we can do is prepare ourselves to adapt or react to whatever is thrown our way. Let's call this dispositional orientation toward the weather: "accepting it."

It wouldn't take much for us to adopt such an attitude, to be somewhat fatalistic about what comes our way. We could just accept weather as a fact of the world, much like we might accept other facts about ourselves: our height, our size, our parentage, our wealth, our intellect, our biological sex, the color of our skin, and so on. Indeed, there are actually a lot of facts about who we are that we might just accept as immutable and unchanging. I stopped growing at about five foot five inches (1.6 meters), and though I would like to be a bit taller, I just can't do anything about it, so I have to live with it. Weather could be something like this. In this way, we might view weather as just a *neutral fact of life*. Neither good nor bad, just kind of – there.

There's something very Zen about this. Why fight what you can't control? We must learn to live peaceably with the land and the weather, changing farming practices to suit the conditions. In arid regions, perhaps it would be better for us not to build our homes or try to plant crops. In colder regions, we might seek to grow only hardy crops, or to raise animals for food. We might aim to live at a higher elevation in the summer months, using ice and cold to keep food from spoiling.

Likewise, there's something very stoic about it too. Stoics like Euripides, Epictetus, and Marcus Aurelius all reflect a resolve not just about the weather but about outward events in general. "Be not angry with outward events, for they care nothing for it," says Euripides. "Some things are within our power, while others are not," nods Epictetus. Marcus Aurelius adds to the insight, suggesting that, "If thou art pained by any external thing, it is not this that disturbs thee, but thy own judgment about it. And it is in thy power to wipe out this judgment now."[2] Here he finds power in embracing the world as it is, accepting it, and allowing that our power lies elsewhere.

My students sometimes also hold attitudes like this, frustrated as they are with the doldrums of industrial civilization. They assert with some confidence that the world would be so much more peaceable and sustainable were we to turn off our screens, put down our technologies, and return to the wild. I typically remind them – romantics and naive primitivists that they are – that we are holding class in an air-conditioned classroom, covered by a roof made of steel and concrete, protected from wind and cold by the engineers and architects who have designed our very large, quite fancy building. I hate to be a buzzkill, but their attitude of acceptance is made possible in part by the technologies that otherwise hum invisibly in the background.

So this brings up the central observation about weather that I make in this chapter and explains at least part of the reason that I've chosen to focus on dispositions as opposed to moral valences (which are, again, a blanket evaluation of whether weather is good or bad, positive or negative). So many of our attitudes about the weather are mediated by the technologies that we have available to us. It is easy to romanticize or think positively about weather without factoring in the extent to which it both challenges us and situates us in the world, but also how the moral valences of weather's impacts are not a simple matter of saying "storm = bad" or "sun = good." Where so far we've discussed the extent to which weather is a force, and a chaotic force that impacts us in substantial ways, it is also shaped by our own interactions with it. It may be helpful, then, to reflect briefly on the ways in which our weather-related technologies have developed and influenced our thinking about our place in the world. What I suggest is that as we move through various stages of development, as we have developed new technologies and learned more about our world, we have in different ways adopted stances toward the world that have influenced our thinking about the moral valence of weather. This is a reflexive arrangement: our stances shift with the technologies, all at once enabling us to let go of past attitudes but also pushing new attitudes in their place.

****

The first agricultural revolution (the Neolithic revolution, about 12,000 years ago) came on somewhat slowly at first, likely by accident. Anthropologists surmise that hunter-gatherer types would collect fruits and berries to eat near their temporary campsites, only inadvertently to drop seeds nearby. Over time, those seeds would germinate and the members of those communities eventually learned that these seeds could be planted deliberately. (It's not clear to me why they wouldn't just be able to observe plants and seeds in various states of growth, as even a basic observer of nature is able to do; but I'm no anthropologist so I can't say why the theory of accident is more persuasive than another theory, of which there are many.)[3] It was quite a bit later, allegedly, that those same people discovered that crops would grow more successfully if they received irrigation. This transition to irrigation is believed to have occurred around 6000 BC in Egypt and Mesopotamia. The Nile, the Tigris, and the Euphrates rivers would all flood during monsoon season, and locals in these regions discovered that they could divert water into their fields. This began first as a system of canals but gradually resulted in more sophisticated methods of irrigation.

Up until that point, people had been primarily migratory, essentially following the weather and setting up camp in hospitable climates. Small communities of people would move to areas where there was ample rainfall; or perhaps they might abandon desertified regions, seeking places where rain was more plentiful, perhaps feeding a nearby river, lake, or pond. They may, alternatively, have moved to coastal regions to take advantage of the cooler weather on the coast, where they could hunt, fish, and cook. Indeed, much of the prehistory of humanity can be traced back to efforts to chase, and accept, the weather. The migration of whole peoples across regions was neither an attempt to resist the weather nor an attempt to harness the weather but rather an attempt to find suitable conditions in which to thrive. It must've been quite difficult to know, without advanced record-keeping, how food and shelter would persist over time.

At the same time that technological innovations in agriculture civilized folks and enabled them to settle down, however, these innovations also introduced a new kind of dependence on nature. Rather than being dependent upon weather more broadly speaking, their newfound rootedness made them dependent upon the agricultural spaces themselves, upon regional food productivity, upon the weather systems localized to regions in which they had developed their technologies. It thus became critically important to find fertile and arable land that people could cultivate, and this meant, in many respects, finding the weather equivalent of a Goldilocks zone: an area that was not too hot, not too cold, not to wet, not too dry, but just right. In these kinds of contexts the idea of "territory" becomes much more important, or at least takes on a different political shape. And though I cannot speak to the anthropology or history of the idea, it is not unreasonable to surmise that primeval notions of "property" emerge precisely from contexts such as these.

The Indus Valley (now Pakistan and India) and the Fertile Crescent (now the Middle East) provided a perfect natural location for primitive farming practices to flourish, thereby freeing people in these regions to engage in more creative endeavors, to trade and barter with one another, and to build sturdier and more significant dwellings. As cultural practices shifted from foraging to farming communities, villages and cities sprung up that bound themselves to the weather of the region. Our earliest techniques for agriculture very much involved accepting the natural environment and living harmoniously with it, even if there were aspects of the first agricultural revolution that also involved learning to redirect and control weather and other natural resources.

Now then, the first agricultural revolution was in many respects still a relatively low-tech revolution. More than anything, however, it helps illustrate how our technologies influence our comportment to the world around us, all at once providing us with previously out-of-reach power to control our environment while tying us more than ever to a given system. In a world absent agricultural technologies, weather looks to be fickle and unpredictable; we are dependent upon its cooperation with our objectives. As we have developed new technologies, weather has, on the one hand, become less worrisome; on the other hand, it has bound us in new ways to our newly emergent objectives.

With the advent of agriculture, humans were able to control at least some of the factors that had kept them nomadic. Their innovations, technical and practical, also allowed them to live harmoniously with the weather and to take advantage of what it had to offer. Waiting for monsoon season involves adopting an attitude of accepting weather, whereas building canals and irrigation systems more aptly involves an attitude of harnessing the weather. But it is critical to see that the orientations themselves are carried by civilizational developments, and that, for instance, monsoons may not have been a particularly big deal when people were more migratory, but that they became a big deal once they were dependent upon the monsoon. A similar story could be told about wildfire. Where thousands of years ago a forest fire may have blasted across a region and, for a short time, forced game out of the forest and into the waiting spears of hunters, now more than ever they pose a substantial threat to suburban homeowners. All this is to say, these orientations, and the values embedded in different aspects of the weather, fluctuate as we live in and with weather.

Resisting It

Another attitude one can adopt toward the weather aims to resist it, or at least to prevent the negative impacts of weather from affecting us so harshly. Our disposition to resist the weather comes at us early in life, as survival tip 101, taught to scouts and campers everywhere: if you find yourself in unfamiliar or hostile circumstances (e.g., lost in the woods, threatened by wildlife, chased by brigands, battered by weather, or pursued by dragons), seek shelter. If you are lost in the snowy north, build a snow cave. If lost in the sunny south, fashion a wickiup. If ejected from your country due to strife, find shelter in a refugee encampment. The shelter needn't be anything fancy. It can be a cabin or a bivouac or even an outcropping of rock. The objective is just to get your vulnerable body out of the way of the various hazards that the world throws your direction.

The imperative to seek shelter is in some respect primordial. Before it is articulated to children in the form of tips or rules, it is therefore "noncognitive," meaning that we don't need knowledge of it to understand it instinctively. It's no different in this way than the imperative that many living creatures themselves understand intuitively. Wolves seek shelter in dens, squirrels in middens, birds build nests in trees that are protected from the cold, beavers build lodges, rabbits live in burrows, and so on. There's nothing particularly technologically or intellectually sophisticated about seeking shelter. People and animals seek shelter because it protects them from the elements. The hazards that the elements present are themselves felt, rather than cognized. (This is important, by the way, because we don't treat or think of hazards quite like we used to, but I get to that point later in the chapter.)

Our capacity to protect ourselves from weather doesn't stop at shelter, of course. We don additional layers of exterior protection in the form of clothing too, and we've done so for millennia. Many versions of the Adam and Eve tale have the pair frolicking *au naturel*, unaware and unconcerned of their nudity until they eat from the tree of knowledge. The story suggests that Adam and Eve took care to cover their parts with fig leaves, but presumably out of modesty rather than protection. Was the Garden of Eden also devoid of weather, much like Olympus? It would be hard to square such a thought with the ostensible abundance of plant and animal life in paradise, but mythology leaves a lot to the imagination.

Less mythologically, there are cave paintings dating back 30,000 years, to the Paleolithic period, that depict humans in skins, furs, and plant matter, presumably more to shield and protect early humans from the elements than to cover up any private bits or to indicate our social station. Textiles developed shortly after that, approximately 27,000 years ago, and with them the idea to make them pretty.

Box 4 Ventilation is the New Black

Bedouin Arabs are sometimes said to wear all black because it prevents them from overheating, though it turns out that it is the physical nature of their robes, and not so much the color of the fabric, that likely contributes to the cooling.[4]

Our improvements in clothing have been both incremental and dramatic. Socks alone are a massive advancement in the comfort department. In the early days of sock development, roughly 5000 BC, socks basically doubled as shoes – outside feet – and were made from animal skins and lined with straw or grass. It was only in the eighth century BC that Greeks started felting socks from animal hairs and wearing them with their sandals.[5] These "piloi," Hesiod advised, should be worn between well-attached sandals and snug to the feet:

> Lace on your feet close-fitting boots of the hide of a slaughtered ox, thickly lined with felt inside. And when the season of frost comes on, stitch together skins of firstling kids with ox-sinew, to put over your back and to keep off the rain. On your head above wear a shaped cap of felt to keep your ears from getting wet, for the dawn is chill when Boreas has once made his onslaught, and at dawn a fruitful mist is spread over the earth from starry heaven upon the fields of blessed men: it is drawn from the ever-flowing rivers and is raised high above the earth by windstorm, and sometimes it turns to rain towards evening, and sometimes to wind when Thracian Boreas huddles the thick clouds. Finish your work and return home ahead of him, and do not let the dark cloud from heaven wrap round you and make your body clammy and soak your clothes. Avoid it; for this is the hardest month, wintry, hard for sheep and hard for men.[6]

Fortunately for school children everywhere, socks are no longer darned from the scratchy, itchy animal wool that was common even just fifty years ago. When I was a kid, my mom used to wrap my feet in plastic bread bags before cinching them into my winter boots. Ostensibly this not only helped squeeze our feet into tight spaces but also kept them dry in case the snow leaked through. We now have silky smooth cotton, nylon, and synthetics that wick moisture away from our feet. Living with and playing in the weather is just so much more comfortable now than ever before. My great grandfather Maynard used to have a saying: "Any old dummy can go out in the wilderness and be miserable. It takes a little planning and ingenuity to have a good time."

On a much larger scale we have, of course, innovated well beyond simple shelters and body coverings. As far back as 2000 BC, the Chinese king Da Yu – Yu the Great – rose in popularity in part due to his efforts to control the floods of the Yellow River.[7] Because there is no written evidence dating this far back, many conjecture that this is just a myth, as Yu the Great eventually became elevated to a water deity in Daoism and many Chinese folk religions.[8] Much later, but still a long time ago, Li Bing in 300 BC built one of the first eastern dams on the Minjiang, a tributary of the Yangtze river – which was later used to irrigate crops to support Dujiangyan city. Since then, weather resistance technologies have proliferated, with dams, weirs, dykes, culverts, diversion canals, groundwater replenishment, levees, bunds, reservoir, wetland areas (sloughs), and even just plain old sandbags. Research suggests that weather-related losses are 98 percent lower today than they were just a century ago.[9]

Box 5 Umbrellaology

John Somerville's fantastic 1941 piece "Umbrellaology, or, Methodology in Social Science" invents the absurd science of umbrella studies – Umbrellalology – to challenge readers to question the nature and specialness of science. He notes that his hypothetical science of umbrella studies is based on observational facts, verifiable predictions, and general laws just like other sciences. Shouldn't it too be considered important?[10]

We've developed tools and technologies that help us better resist the weather. The umbrella technology I mentioned at the beginning of this chapter is in some ways worth the "oohs" and "ahs" that it elicited from the kids at the bus stop. Not only does it prevent us from getting soaked, but it is pretty cool to see umbrellas magically fold and unfold their fabric shield. In other ways, it's a fairly simple technology, hardly even creative, notable perhaps because of its simplicity. There are images in Egyptian hieroglyphs of people using devices similar to umbrellas to shield themselves from weather, but some date the invention of the umbrella back to China, 3,000 years ago.[11] And the same can really be said for many of our modern protective technologies, that they're just slight improvements on a theme: shirts, shoes, hats, jackets, carriages, tires, roofing, tarps, tents, refrigeration, air conditioning, etc. We've even developed mostly invisible protections in the form of sunscreen. Our technological capacity to resist weather has grown immensely over the past several thousand years.

*****

The second agricultural revolution came on quite a bit later than the first, in the mid-seventeenth century, long after many of these innovations in Asia, and just as the Enlightenment in Europe was kicking into high gear. Sometimes called the "British" agricultural revolution, it involved mostly improvements in knowledge and innovations in policy. Farmers learned improved crop rotation, leaving the land fallow to allow it to rejuvenate while different crops could be grown in nearby farms. Commons areas were divided up into private ownership rights and the principle of "enclosure" was introduced, making larger farms possible. Market barriers to exchange were removed, freeing farmers to set prices differently. Transportation infrastructure was improved, providing channels through which crops could be distributed to wider consumer markets. Sure, there were some new technologies that came on the scene as well, such as more efficient plows that could cut through dirt without as many domesticated animals or modifications to field drainage systems, but many of these innovations were more conceptual than physical.

Throughout this period, farmers introduced conceptual innovations to limit the hazardousness of weather, pooling resources with other farmers or within their community. If crops were to fail due to blight or storm, a sufficiently large collection of peasants, serfs, farmers, and shepherds could help soften the blow by stockpiling resources in different locations and redistributing them in cases of need. Clever merchants were likewise able to devise strategies for coping with the variability of weather. For instance, rather than shipping their entire collection of goods in a single ship across the stormy seas, they would instead spread their cargo across many ships to prevent loss.

There are many strategies that one might deploy to resist the weather: one strategy is just simply to *retreat*, to move to higher ground; another is to become more *resilient* (i.e., to strengthen our ramparts by innovating and developing); a third strategy is to *adapt* (possibly importing technologies or techniques from elsewhere); a fourth strategy is to *pool risk* (to diversify one's portfolio so the impacts are less heavily felt). There are also many reasons to develop or adopt technologies and innovations, many of which we covered in Chapter 2: to reduce the impact of weather damage, to protect private property, to save lives, to prevent health effects, to protect commerce, etc. – essentially to stop the destructive impacts of weather. But the central reason to resist the weather is that sometimes weather poses a *hazard* to us. The idea of innovation is to avoid these hazards by developing technologies and interventions to prevent injury or harm. That may sound obvious, but it is actually far less clear than it appears.

Where many modern policy economists might construe these weather hazards as "risks," the idea of risk – intimately bound up as it is with mathematical conceptions of probability – only entered our lexicon very recently.[12] Rather, in the past, those living with nature had to accept the slings and arrows that weather tossed their direction, but they could resist the weather by devising clever workarounds. I return to this later in the chapter.

Harnessing It

There is, of course, a third option here, that of grabbing the reins of nature and harnessing the weather to direct its output in a more positive direction. In some cases we've built dams and levees to protect ourselves from weather, but in other cases we use them to help us achieve our objectives. Sometimes they serve both purposes at once.

Early forms of water collection involved pooling water into lakes or reservoirs. Rainfall is known to be inconsistent, so agricultural populations across the Fertile Crescent in Mesopotamia came up with the brilliant idea of collecting water and saving it for later. They would build not only berms to protect against flooding but also reservoirs to capture runoff that could later be used to irrigate crops.

Most readers are likely familiar with the water wells that have been used all across the globe to tap into naturally existing aquifers. The image is quaint, of Snow White wistfully wishing at a well, essentially a pit with a bucket on a rope. But technologies for finding water have been around for centuries. The *qanats* of ancient Iran and elsewhere throughout North Africa and the Middle East were developed thousands of years ago to transport water through the desert and irrigate cotton fields. Meanwhile, across the globe, the Peruvian technology of the Nazca peoples (200–650 AD) created *puquios* aqueducts. These were long, manmade, spiraling stone tunnels – made up of "ojos" or eyes – that served as access points for farmers to reach underground water reservoirs. Acequia communities, originally emerging in Spain, used water above ground, in long ditches, essentially serving as a public water supply or what we might call in more contemporary parlance a "common pool resource." Communities built entire systems of governance around the idea that water could be shared.

Box 6 The Power of Love

The average bolt of lightning carries 10 gigawatts of electricity. Marty McFly and Doc Brown needed to capture just 1.21 gigawatts from the clocktower to send Marty back in time. Had they instead tried to use solar panels, they would have needed almost 1.9 million of them.

We've used wind as well. Certainly sailors have historically harnessed the wind to pull their ships. Farmers have been using wind technology to pump water or to mill grain into flour for at least 3,000 years. More recently, in 1887, inventor, entrepreneur, and professor of natural philosophy at Anderson's College in Glasgow, James Blyth, was the first to attach an electrical generator to a windmill, and by 1888, millionaire Charles F. Brush – innovator of the carbon arc lamp and founder of the Brush Electric Company – had designed and built an enormous fifty-six-foot wind turbine on his property. He used this turbine for over twenty years to power his mansion – all before the turn of the twentieth century.[13]

Nowadays we harness the weather in much more familiar and somewhat more fantastical ways. Many of us have solar panels on our homes and use the sun to generate electricity. Giant wind turbines dot the landscape of many countries and coastlines. The same goes for water power, which is highly contingent upon rainfall. Some companies have even tried to harvest lightning, recognizing the tremendous power contained within.[14]

When I was in graduate school, at one point I worked on the Central Arizona Project, which was at the time one of the largest water projects in the world. Essentially a 336-mile-long aqueduct that moves water from the upper basin of the Colorado River, where I live now, to lower-basin states like Arizona, all by way of California. Most of the Mountain West is desert of some sort or other, and agricultural and economic growth has been historically dependent upon precipitation. There's a regional mantra out here – "Whiskey's for drinking, water's for fighting" – that neatly encapsulates the tension over water and rainfall. (Don't even get me started on the Colorado River Compact, which is a legally binding document that allocates 16.3 million acre feet of water to eight states and 1.5 million acre feet to Mexico, but that was negotiated during a particularly flush set of years for rain, such that the annual flow of the Colorado River was estimated to flow at 16 million acre feet. Subsequent years have shown that the average flow rate is closer to 12.4 million acre feet. Guess which downstream country doesn't get water?)

The Central Arizona Project has since developed in a new direction, as many crops grown in Arizona, cotton in particular, but also wheat, are now owned and operated by investors far away, in other parts of the world. Most recently, heavy investment has come from Saudi Arabia. This means that much of the water that moves through the Colorado River eventually lands in Arizona, and ends up in Saudi Arabia, rather in the United States. This is yet another way in which our control of the weather has redistributed the precipitation that once was the province of the gods.[15]

******

The third agricultural revolution – the so-called Green Revolution (1940–1970) – was as much an outgrowth of the orientation toward harnessing nature as anything else. It largely involved the cultivation of high-yielding crops, the development of chemical fertilizers and pesticides that could mimic the powers of nature, the introduction of mechanized farming with heavy machinery, the cultivation of monoculture crops (largely for economic reasons), and of course continued advancements in irrigation technologies. In this respect, it was mostly a revolution of scale and scope, where economic factories industrialized agriculture.

The dominant dispositional orientation here aims to harness the weather by finding ever more efficient ways to take full advantage of whatever the weather had to offer. Modifying tomatoes to grow in harsher climates is a way of harnessing the weather.

In recent years, policymakers and environmentalists have sought to capture the value-conferring dimensions of nature in terms of "ecosystem services," pointing out that it is easy to overlook the wide array of benefits that nature provides because these benefits in many cases mostly happen in the background. Attributable in part to work by Paul Ehrlich, Gretchen Daily, and Robert Costanza (among many others), the idea is not just to call attention to such services but to put a price or a value on these services so that they can be better factored into cost–benefit assessments.[16]

The USDA defines ecosystem services as "the direct and indirect benefits that ecosystems provide to humans."[17] They then separate the services of nature into four primary categories:

- provisioning services
- regulating services
- supporting services
- cultural services

Provisioning services point to the ways in which nature provides for us. In our discussion of weather, these services might come in the form of fresh water, bright sun, cooling cloud cover, gentle breezes, and so on. *Regulating services* serve to balance out ecosystem processes, so weather may help evaporate water to purify it, help distribute nutrients through soil, pass along pollen to fertilize other neighboring plants, distribute seeds through the wind, thin forests and prevent some species from proliferating, and so on. *Supporting services* maintain processes that are already in place, supporting stable conditions for wildlife and plants to persist. Weather assists with nutrient cycling, for instance. And finally, *cultural services* are meant to capture all of the nonmaterial benefits that nature provides. And in our case, this can include providing snow for skiing, sun for sunbathing, warmth for harvest festivals, and so on.

These categories, in some form, appear in multiple other locations as well. The Millennium Ecosystem Assessment (which is an extremely important and influential document created by the United Nations to assess the "consequences of ecosystem change for human well-being") utilizes the ecosystem framework heavily.[18] Inasmuch as the emphasis of the assessment is not only on change, but also on well-being, it aims to provide a scientific basis for action. So too with UN initiatives like TEEB (The Economics of Ecosystems and Biodiversity), which aims to make visible the interplay between economic and ecological factors through valuation.[19]

The key distinction here is between the functional and the material. Specifically, the ecosystem services literature aims to call attention to the value-generative capacity of ecosystems ("natural capital") that otherwise gets swallowed up by an emphasis on resources or goods themselves.

To illustrate: There are lots of *material* goods in ecosystems – water, timber, produce, meat, fish, minerals, metals – that we have, for centuries, readily recognized and utilized. We've basically viewed most of the world around us as a repository of raw materials that we can piece together to create goods. We can find things in the world, pluck them out of their environment, and use them to make our lives better. It is easy to think about the material stuff in our environment only as resources because the resources of nature are easily (or relatively easily) marketized.

Ecosystem services, by contrast, are more *functional*: They help create and support these resources, much like a factory might function to create the goods that emerge from an assembly line. In this way, ecosystem services are determined by *configurations* of stuff in the world that serve a particular role or function in ensuring that those raw materials are available to us. That is, they're neither stocks of goods nor flows of goods but *functional relationships between the material of nature*. The world has to be structured in a particular way to have these functions work. Again, to illustrate: we can have a bunch of gears, valves, pistons, cylinders, and belts sitting on the table in front of us, but it takes putting these items together in a particular way to make an engine. Once we have the engine, however, the engine can perform a function for us.

As one might imagine, weather is a critical piece of this discussion. A string of sunny days provides heat that evaporates water sitting in open ponds that then forms into clouds that then rains back down on Earth. Dust or pollen particles in the air provide a medium for ice crystals to form, which then sprinkle down on the planet as snow, which then slowly melts over the spring to provide clean water through the summer. Heat, pressure, and moisture serve to create wind that strips pollen from stamens and pollinates plants. That is, weather and the forces of nature do things for us. Humans don't just need a steady supply of water, we need a steady supply of *clean* water. Likewise, we need a relatively stable climate upon which we can depend, a constant stream of sunlight, and warm not freezing temperatures, among many other things.

As Western capitalism has proliferated and taken to transforming natural resources into value and wealth, businesses have largely ignored the beneficial services of nature and not factored them into their production calculations. They are, essentially, getting these services for free. Sometimes, if not often, when companies focus on their production but fail to recognize the services they get for free, they mistake "free" for "given" and take them for granted. They may, in virtue of this, pollute or sully or cut down or pave over the systems that provide those services, thereby inadvertently undermining either their own productivity or the productivity of other industries that are also dependent upon those services. If a factory emits soot and ash into the air that later rains down on a nearby town, weather would wash this pollutant away, ostensibly clearing the air but harming the population or plant and animal life on the ground – indeed, this is precisely the concern with acid rain – that the weather systems will self-organize and wash out pollutants, all the while creating havoc on the ground.

The shift from environmental goods to ecosystem services may seem like an innocent conceptual development, simply an effort to call attention to benefits that otherwise go unnoticed, but consider how abstract and actuarial this move is, how it separates us from the world around us; and in our particular case, from the experience of the weather, from the phenomenon of being buffeted by winds or having our family photos dissolve in a flood. Talking about weather in terms of its services shifts our perspective on weather into a functional relationship, essentially transforming weather into an instrument or a tool for us.

The ecosystem services stance is therefore more than just an academic observation about the usefulness of weather. It reorients our thinking about weather and about ecosystems in general. It is an intellectual innovation, a kind of technology that mediates our thinking about the world. It is shot through with value suppositions that, at base, propose that we understand our relationship to nature primarily in cost–benefit terms. Like many of the technologies that we have been discussing, it is a morally laden endeavor, aimed not only at identifying facts about the weather and environment around us but also at recasting the moral valence of nature's services from neutral to good. Epistemically speaking, there is the simple matter of what we know and understand about weather and its impacts on us. Morally speaking, there is the additional unanswered question about what we should do about these impacts.

The ecosystem services stance has been criticized by various theorists on numerous other grounds.[20] Some have suggested that it is anthropocentric. Others have argued that it encourages us to think about our relationship with nature primarily in terms of resources; that it stacks the deck against nature by falling victim to the commodification of nature. Others claim that it is relying upon vague definitions and classifications. Some argue that it runs in conflict with other biodiversity conservation objectives. The arguments include that it also operates along a monotonic scale that is exclusively economic in nature.

Clearly there are reasons to make ecosystem services arguments. Top among these is a pragmatic one: that policy-makers are inclined to be persuaded by cost–benefit analyses, and in theory at least they will be responsive to the financial benefits of these services if they are pointed out. With enough science and know-how, goes the reasoning, the powers that be will accept and value those services. But, as we've seen, weather is a double-edged sword. It can be just as devastating a force as it is productive a force. Just as there are ecosystem services, there are ecosystem *dis*-services. One winter frost, one Category 5 hurricane, one prolonged drought, one blazing wildfire, and whole economies can come crashing down. This productivity or devastation is, on the one hand, a consequence of what channels we have available to us to accept or reject the services on offer and, on the other hand, a consequence of whatever objectives we may already have in play.

What gets further left out of this discussion, unfortunately, is the extent to which these functions can be replaced by other functional technologies. So long as we have an adequate way of fulfilling those functions, say by running water from a nearby lake to irrigate our crops, or by keeping our air conditioners running during the hot summer months, we don't have to worry so much about those services. It would seem from the ecosystem services discussion that the functions that nature offers are always the most efficient and best way of receiving those services. But it should be quite clear to all that many of our technologies, particularly those that are and have been successful at resisting weather, are also in many respects better at providing the services that weather is alleged to provide than weather itself. And this is a particular problem if the objective is to do more than understand what weather is and to think of weather as a replaceable component in a functional system. More on this in Chapter 5.

All this is again to point out that human innovation – physical and conceptual – has helped us over the years adapt and learn to accept the weather; and in turn, that our attitudes about the weather and the impacts of weather are largely contingent upon historical context. Agricultural regions have developed in part because weather is conducive to growing certain crops over others. Some places are highly suited for growing bananas, such as in the tropics; where other places are best suited for growing corn, rice, wheat, or peppers. Potatoes may need the ground to freeze over in order to prevent blight. Rice thrives in heat but needs a ton of water, while wheat grows better in temperate climates. These staple crops and cuisines from different parts of the world are in part in consequence of both climate and weather.

And stories such as these can be told across industries and communities much wider ranging than agriculture:

recreation, transportation, construction, the military. The ski industry is deeply dependent upon snowfall, and whole regional economies are built up around snow and ice. Sunbirds in Arizona, Florida, and California flee the harsh winters of the north to find solace by the warm waters of the ocean. Trucking and shipping gets considerably more difficult as winter weather sets in and otherwise navigable routes have to be avoided in order to transport goods from point A to point B. Major construction projects are thrown off as weather interferes with plans. Even retail businesses tend to do better on warmer days when people are more inclined to shop.[21]

The Social Construction of Weather: Facts, Hazards, and Boons

If the foregoing focus on dispositional orientations makes sense, or at least resonates with your own experiences, then it is important to see the way in which our attitudes toward weather – that is, accepting it, resisting it, or harnessing it – are not themselves set-in-stone dispositional orientations, immovable and characterized by prior value commitments, but also very much shaped by our sociocultural context. They have ebbed and flowed socially, culturally, and politically over the centuries. Having the right tools changes our relationship to weather (as it does to nature more broadly). I suggest that the formation of these attitudes is a *reflexive* process, meaning that it feeds on itself.

Sometimes when I'm teaching environmental ethics I tell a story about the various different moral valences that have characterized our relationship to nature over history. It's not exactly precise, but it's a similar sort of cartoon history that I'm telling here. I suggest that, at times, nature has taken on a negative valence, as with the Puritans, who thought that we ought to protect us from the outside world, who saw in nature threats of demonic possession and witchcraft. I point out that, at different times, nature had a more positive valence, as with the Romantics, who thought that nature should be accepted because it provided peaceful respite from bustling and dirty urban areas. And then I add that nature has, as with the naturalists, taken on a neutral valence, where it was seen is neither good nor bad but rather just value empty.

The dispositional orientations that I carved out earlier mirror the three different valences of nature that I use in my class. Here, however, we're not asking so much what the moral valence of weather is – whether it is good, bad, or neutral – but rather how our attitudes or orientation toward weather shapes our thinking both about the valence of weather and about weather itself.

I've chosen this attitudinal framework because I think attitude-talk more directly calls attention to the contextual basis that informs our thinking about values and impacts, which in turn shapes the social and technical apparatus that we build to respond to the world around us. Plus, values talk is often charged and confusing for those not writing in scholarly philosophy circles. It tends to mean something like "the values we hold" as opposed to "the values that are, *simpliciter*." In Chapter 2, I divided weather into positive and negative impacts, speaking of weather in terms of its destructive power and its productive power. I think that's kind of intuitive. So this discussion of dispositions really just layers on that conversation. As you were reading this you may have thought, "Gosh, yes, but the reason anyone would resist weather in the first place is because they already recognize it as negative." I worry, however, that this masks part of the way in which we come to hold values. Our capacities change how we value the world.

I really want to emphasize, though, as I've tried to do already, that our dispositional orientations are always already mediated by the technologies that are available to us. A torrential rain without an umbrella makes for a nasty trip to the bus stop. A tornado without shelter is a terrifying confrontation with death. A snowstorm without boots, mittens, and hot chocolate is a teeth-chattering nightmare. With them it is a romantic evening.

When we adopt an attitude of accepting the weather – or at least some aspect of the weather – we do so in a way that takes that aspect essentially as just a *fact of the world*. It has a neutral moral valence. When we instead adopt the attitude of resisting the weather, this shifts the moral valence of that aspect of weather and treats it as a kind of *hazard* or *threat*. And so too when we adopt the alternative attitude of harnessing the weather, this shifts the moral valence in the positive direction, treating the moral valence of that aspect of weather as a *boon* or *benefit*. Indeed, this is precisely what the people who have been advancing the ecosystem services argument have tried to do: to get us to see that our attitude of simply accepting the weather and treating it as a neutral fact of the world has been mistaken, and instead we should think of its services a more positive light, as a boon.

The last two orientations, more active than passive, lead us to develop technologies that mediate and thereby further alter our experience of and attitude toward weather. When my son grabbed that umbrella at the beginning of the morning, the simple technology of blocking the rain changed his relationship to the rain. No longer was it an inconvenience that would have made his trip to school a soggy, sodden mess. Instead the umbrella itself was both a curiosity and an inconvenience that he had to hold over his head in order to stay dry. That act of holding the umbrella over his head, though something of a burden for me, also created a dry mini-environment that followed him down the street. Despite the rain outside, it was not raining under the umbrella. He could both be "in the rain" and "dry in the rain." His experience of rain and therefore his thoughts about the value of rain were therefore strongly framed by the technology. Allegedly there are "pluviophiles" in the world – people who love rain and rainy days – but I strongly suspect that their philia is at least partly attributable to the protective technologies they have available to them.

Philosophers of course have a long and sordid history with critiques of technology and specifically with the way in which technology has modified our vantage on the world. From the Jacques Ellul to Lewis Mumford, from Martin Heidegger to Langdon Winner, theorists have expressed worries that technology may be doing more than offering mere utility to us. Ellul's worry was that we are undercutting our freedom in the technological society, whereas Heidegger's worry was that technology serves to frame and mediate our world so much so that we begin to rethink the essence of the world around us.

I don't have space in this text to cover the nuances of their different views, but Heidegger's observations about windmills may be helpful here. He thought, for instance, that by installing hydroelectric plants along the Rhine, we were transforming our thinking about nature, turning nature into what he called "standing reserve":

> [The hydroelectric plant] puts to nature the unreasonable demand that it supply energy which can be extracted and stored as such. But does this not hold true for the old windmill as well? No. Its sails do indeed turn in the wind; they are left entirely to the wind's blowing. But the windmill does not unlock energy from the air currents in order to store it. [...]
>
> The hydroelectric plant is set into the current of the Rhine. It sets the Rhine to supplying its hydraulic pressure, which then sets the turbines turning. This turning sets those machines in motion whose thrust sets going the electric current for which the long distance power station and its network of cables are set up to dispatch electricity. In the context of the interlocking processes pertaining to the orderly disposition of electrical energy, even the Rhine itself appears to be something at our command. The hydroelectric plant is not built into the Rhine River as was the old wooden bridge that joined bank with bank for hundreds of years. Rather, the river is dammed up into the power plant. What the river is now, namely, a water-power supplier, derives from the essence of the power station.[22]

This passage mirrors an idea that we saw earlier with the effort to introduce ecosystem services arguments. That is, that many people think about nature in terms of stocks of materials only, rather than in terms of their function or purpose or idea. Where we might naturally assume this is just a kind of myopia specific to business entrepreneurs, so far as Heidegger is concerned, it's modern technology in particular that changes our relationship to nature.

Our technologies mediate the world around us, changing how we experience it, framing it in a way that informs our decisions. This is a symbiotic, self-reinforcing, reflexive process. Many of the technologies that I mentioned earlier not only shape our thinking but are themselves a consequence of the dispositional orientations we have just discussed: identifying a need or an opportunity and then building out from there.

Couple this with work by sociologists Ulrich Beck and Anthony Giddens on the "Risk Society."[23] They observe that we have built up an enormous industrial apparatus that, on one hand, has made life much less risky for us. We have been able to identify hazards in the world, innovate technological and policy workarounds to address them, and ultimately reduce those risks to us. At the same time that we've done this, however, we have also generated a whole new category of risks, what they call "manufactured risks," that are a consequence of the technologies and innovations themselves. For us, this introduces new vulnerabilities, not unlike the vulnerabilities felt by those in the Fertile Crescent who became more vulnerable to monsoon rains also while depending on them.

Their argument is pinned not so much to the essence of technology and how it frames our thinking – as it is with the other thinkers I mention – but rather to the reflexivity of risk itself and to the nature of risk analysis as a tool for mitigating hazards. The Chernobyl disaster looms large over this work, providing a stark example of how our nuclear infrastructure – a technology created to harness nature and provide a stable source of power in the face of dwindling natural resources – created a brand new type of precarity that itself needed to be contended with. Part of the way Beck and Giddens get there is by differentiating mere hazards – that is to say, pitfalls or dangers in the world – from risks.

For most of human history, we have wandered through life and encountered pitfalls and hazards that have made our lives more difficult. Our technologies have helped us mitigate the harm that might eventuate from these hazards. Hazards are in this way physical threats or dangers to us that obstruct or thwart our ends. Only once we were collectively sophisticated enough to develop a mathematics of probability and statistics were we able to understand hazards in terms of the probabilities that some negative state of affairs would come to pass.[24] In other words, when we ask a question about risk, we ask how probable it is that we may face a given hazard. There may be low risk of flooding or a high risk of wind damage.

The word "hazard" originates from the Arabic term "al-zahr," which is literally the word for "dice." I mentioned briefly in Chapters 1 and 2 that weather is inextricably caught up in luck, and this is no less true about weather itself than it is true about our attitudes toward weather. It is very much the case that our technologies help us buffer the vagaries and impacts of weather, minimizing the likelihood of these hazards, softening the blow, preventing bad stuff from befalling us. To be sure, much of the discussion in Chapter 2 covered the ways in which burdens and benefits have been scattered haphazardly across different communities. (The meaning of the term "haphazard," notably, somewhat redundantly calls attention to the fickleness of hazards. The root word "hap" also suggests something about fortune or luck, as in "hapless," again invoking the omnipresence of the Greek god Kairos and the Roman god Tempus – the gods of luck and weather.) Here it is important to contrast *hazards* with *risks* because, though similar, these two ideas have been in tension from the very beginning.

By reducing the impacts of weather into quantities, by reconceptualizing states of the future as probabilities, we put ourselves in a better position to manipulate and manage the hazards of nature. Risk, ultimately, ties luck or probability back into these threats. In this way, the idea of risk itself serves as an innovation of knowledge that shapes the way that we think about our relation to hazards. (This becomes increasingly important as we move on to discuss meteorology and the science of prediction in Chapter 4.)

******

In this chapter I've looked at the various mechanisms and interventions that we've deployed to try to protect ourselves from, or mitigate the impacts of, weather. I've suggested that as we've made practical and technical improvements to prevent hazards from befalling us, we have also reoriented ourselves and shifted our take on its moral valences.

Though I have laid these three categories out in an order so that each agricultural revolution is linked structurally with a dispositional orientation, please don't make the mistake of thinking that these revolutions are meant to track the orientations in any causal or characteristic way. There are aspects of each of these dispositional orientations throughout history – sometimes we accept or embrace the weather, other times we resist it, and still other times we harness it. Rather, the dispositional orientations are a way of thinking about how we relate to the forces of nature, and specifically to the chaotic force that is weather. More critically, these orientations point to the ways in which our thinking about the goods and bads of weather, the facts, hazards, and boons of weather, have been challenged and shifted over time. It is my view throughout this book that weather is always and forever a volatile background force that serves as a counterweight to our own collective or individual efforts to civilize ourselves, to build a civilization. Weather has always been a part of that effort, by both giving and taking away.

On one way of looking at things, talking about agricultural revolutions trots out the slow march of scientific progress. But on another way of looking at things, revolutions also mark substantive shifts in our comportment to the world, as our technologies have increasingly come to mediate our relationship with nature.

So let's return a third time our working definition and see what we can add:

Weather is a destructive and productive force that, *coupled with human creativity, is an engine of change and innovation.* Weather shapes how we take up the world and how we take up each other. It is a force that can be accepted, resisted, or harnessed; a force that gives us a reason to develop and innovate.

Necessity is the mother of invention, they say, and weather is the mother of necessity. What I've tried to emphasize in this chapter is that weather is both productive and destructive, true, but its valence largely depends on how we choose to live with it, acknowledging all the while that weather is a force that we do not and cannot control, at least not in a direct sense – yet. And that's precisely where we should now turn: to the question of understanding the weather and anticipating the likelihood of different weather events coming our way.

4   Seeing through the Fog
Predicting the Weather

I lived in Tucson, Arizona for two years, from 1995 until 1997. The local weatherman there – a man named Michael Goodrich (aka the "Weather Wrangler"), regionally famous for his dry humor and deadpan weather gags[1] – appeared to be one of the least enthusiastic, but possibly the quirkiest, people in broadcasting. For nine months out of the year, he'd offer the same daily forecast: "Hot." The remaining three months he'd suggest that the weather would be "Warm," with the caution that periodically temperatures would dip into the mid-forties (which is about seven degrees Celsius). Every now and then we'd get a rainstorm, and when we did, he'd pull out popcorn, glitter, party hats, and noisemakers to try to bring some festivity to his monotonous job. The gag, of course, was that the weather in the Sonoran Desert doesn't change much, and tasked with forecasting the weather, there's nothing for a meteorologist to do. But he was also at the same time making light of a much wider national trend to disregard meteorology as pseudoscience and to think of meteorologists as the court jesters of the news industry.

Perhaps better depicted by the movie *Groundhog Day* (1993) – which is about a hapless and bored weatherman who gets stuck in a time loop, repeatedly returning to the same day, on the same morning, with the same people, offering the same bromides, essentially farming his expertise out to a rodent[2] – local news weathermen are often the butt of jokes.[3] They are portrayed as fake scientists who have little to contribute to the news apart from a thirty-second sound bite about the most boring and wishy-washy topic of all. Some of the derision is well deserved, of course. Weathermen are wrong a lot, and they were certainly a lot more wrong many years ago. Forecasting the weather has never been a particularly reliable science. Any child of the sixties, seventies, or eighties who stayed up late listening to the weather forecast, dreaming of an impending snow day off from school, knows all too well the deflation one can feel upon waking to see dry ground. Despite this, there's actually a sophisticated and highly technical science behind their work, with thousands of years of history.

The modern science of weather prediction is relatively new, but the practice of weather prediction dates back centuries. There is, of course, the superficial observation that even a caveman could look out the mouth of his cave, see a rainstorm or a blizzard, and naturally ask himself what tomorrow holds. But historical records suggest that whole populations of people were looking to the skies in an effort to understand what the weather would be.

The Inca, for instance, believed the sun and the moon to be gods with control over the weather. They worshipped the stars and constellations, identifying many individual stars as connected to individual animals. Their early meteorological observations were tied to buildings and monuments that, if viewed across the horizon, would give indications of when to plant. According to Inca lore, depending on how the star cluster Pleiades appeared in the sky – whether bright and large, or dim and small – it was thought to give an indication of when the rain will fall. And as insane as that sounds, modern meteorological techniques have suggested that it turns out to be not such a terrible way to know when to plant.[4]

In Hawaiian mythology, the crescent moon is said to be either "wet" or "dry" depending on whether its horns are facing up or down. A crescent moon with horns up, it is believed, looks like a bowl that will fill with water, and thus was said to capture the water.[5] When the horns are down, it looks to be spilling water out. Likewise, in early Israel, the horns of the crescent moon were said to be of the bull or the moon god, which was likely associated with storms, and depicted as such in early Near Eastern art.[6]

Meteorology in Antiquity

In the late Shang Dynasty in ancient China (roughly 1250–1050 BC), farmers engaged in the practice of divining weather events using "oracle bones" – pieces of ox scapula or turtle plastron (which is the underside shell of the turtle). Though specialists in *plastromancy* were mostly limited to China, the practice of *scapulomancy* was more widespread, occurring in the Americas (among the Cree and Innu peoples), Mongolia, Greece, Serbia, and South Africa, though less for weather-related purposes and more to anticipate future events. There were also, amazingly, other forms of divination as well, including *sternomancy* (mostly a Greek practice that involved making sense of bumps on the breast bone); *hippomancy* (which involved divining weather events by watching the behavior of horses); *ornithomancy* (watching the birds for omens); *hepatomancy* (divining the future by examining the entrails of animals, specifically the liver); and *anthropomancy* (doing the same, but with human entrails); not to mention *geomancy* (looking at sand), *oneiromancy* (examining dreams), and *necromancy* (communicating with the dead). There sure were lots of fancy -mancies!

Aphorisms in meteorology were also common. The poet Hesiod, in *Works and Days* (700 BC), offers what is essentially one of the first-known farmer's almanacs in Western civilization, replete with quippy lines framed like universal truths. There he links astronomical details with weather events, reflecting what was at the time a suspicion about the nature of weather: that it was a consequence of celestial bodies. (Hence the term "meteorology" to describe terrestrial weather events. In its early days, the practice of meteorology was thought to relate to meteors – or metéōron [μετέωρον], which are "high-up things" – and so the naming convention makes sense.)

In about 400–300 BC, the civilizations of the world began thinking more analytically about weather. Chinese thinkers began cataloging the movement of the sun and devised one of the first calendars. It was during this period that Aristotle published his *Meteorologica* (c. 340 BC), which ended up being one of the most important and comprehensive works on weather and persisted in its influence up until the Enlightenment.

Aristotle wrote on basically everything meteorological in this text, and even had a theory of rainbows.[7]

It would be mostly unfamiliar to us, who now understand rainbows to be the result of the diffraction of light through droplets of water. His idea was not diffraction, but rather reflection, that rays from our eyes were directed toward the clouds and then deflected back toward the sun, as if we were looking in a fluffy cloud-like mirror. The eyes *create* the rainbows – they don't *perceive* light passing through clouds but instead project rays onto surfaces and, in the case of rainbows, this light is reflected back to them.

Box 7 The Rainbow Connection

Renowned philosopher, scientist, and alchemist of the Middle Ages, St. Albertus Magnus (c. 1200–1280) – known to many as *Doctor universalis* and *Doctor expertus* (a name befitting a supervillain if there ever was one) – was one of the first to observe that raindrops take the shape of tiny spheres.

He speculated on this basis that this is how rainbows are formed, though he was incorrect about the mechanism through which drops form rainbows. His conjecture had less to do with Aristotle's theory and more to do with the idea that colors are produced through a layering of drops, something like paint in the sky.

In another important text, *Politics*, Aristotle writes of the philosopher Thales of Miletus (624–546 BC), who was widely known to be an intellectual sort, which I think meant that he was viewed by others as a kind of ancient Greek nerd.[8] As Aristotle tells it, Thales was often accused of being a useless philosopher, of being one who knew little of the world and who had, as a consequence, no business sense. Many taunted him for his intellectual pursuits, suggesting that his efforts were pointless. At some point Thales grew frustrated with the taunting and teasing. Determined to prove his hecklers wrong, he applied his knowledge of cyclical weather patterns to business and anticipated a robust olive crop later in the year. With but a little money to his name, during a cold winter, he bid on and purchased all of the olive presses in Miletus and Chios. His wisdom about weather patterns, about demand, and about monopolies, made it possible for him to become a wealthy man, proving his critics wrong. Apart from this story, Thales is widely attributed with advancing the idea that rain is caused by evaporation from the seas.

In *On the Heavens* Aristotle posited that there were four primary elements: fire, air, earth, and water. Each of these elements could be described as having these contrary qualities: cold/hot and wet/dry. Earth was cold and dry, water cold and wet, air hot and wet, and fire hot and dry.[9] He thought, for instance, that when water is boiled, it evaporates and becomes air, which he described as a shift from the "cold and wet" to the "hot and wet." Some of his conjectures are plausible, particularly when you consider his rationale. This was the precursor, notably, to the humorism of Hippocrates that I covered briefly in our discussion of environmental determinism (where, remember, the elements of the body can be divided into four humors: phlegm, blood, yellow bile, and black bile).

Aristotle was by this time already railing against the atomistic theory of the philosopher Democritus (460–370 BC), who preceded him by a few generations and who also had a predilection for forecasting the weather. Though Democritus's "theory of small particles" was popular among thinkers of the time, the Aristotelian theory of four elements – almost laughably retrograde and simplistic now – slowly came to replace Democritus's worldview. Notably, it was not the smallness of Democritus's particles that so bothered Aristotle. Rather, Aristotle was hostile to the notion of a "void," which was essential to Democritus's theory, and in particular he objected that the void was impossible. "Nature abhors a vacuum," Aristotle is often quoted as having said – which he meant in its truest sense, before it was co-opted by Henry David Thoreau[10] and George Bernard Shaw[11] – meaning literally that anywhere a vacuum might've been given to develop, some of these four elements would swoop in and fill the gap. Instead, he thought that the universe was composed of a "plenum," which he surmised to be squishy and "springy," like globules in a bubble lamp, pushed and pulled by the forces of gravity and levity. So the four elements were sort of sloshing around in the atmosphere, pushing and tugging one another, and that's how we get weather.

Shortly after Aristotle finished his *Meteorologica*, his student, Theophrastus of Eresus (372–287 BC), wrote a book on weather forecasting called *De Signis Tempestatum* (On Weather Signs).[12] This book offered a bunch more aphorisms, proverbs, and sayings that would guide meteorology for the next 2000 years:

- "If the winds are from the South, snuff on a lamp's wick signals rain ... and if it is finally granulated like millet seeds and shiny it signals both rain and wind" (p. 67).
- "Whenever sheep or goats copulate, this is a sign of a long winter" (p. 75).

- "There are also signs on the sun and the moon: black ones are signs of rain, red ones are signs of wind. If the moon stands straight when a north wind blows, westerlies are accustomed to follow and the month continues to be stormy" (p. 75).
- "A dog rolling on the ground signals a great amount of wind" (p. 77).

It's not the most impressive bit of advice, but it sure makes for entertaining reading.

Though surely it was the case that people were looking for signs long before Theophrastus, it was disciplinization of weather forecasting that really took off and persisted through to the Middle Ages. Hipparchus of Alexandria (190–120 BC) allegedly created a calendar of astronomical events that was supposed to anticipate weather conditions. The famous astronomer Ptolemy (90–168 AD) continued this tradition in his work *Phases of the Fixed Stars and Collection of Weather Signs*, in which he added mathematical and geometrical meat to the bones of the astronomical musings of Hipparchus. Nobody ever stopped thinking or writing about the weather, but they did it mostly through an Aristotelian lens: fire, water, air, earth; hot, cold, wet, dry.

Many throughout antiquity were excessively focused on reading and understanding the winds. In the Hebrew Bible there were four alleged directions of the winds, one coming from each cardinal direction: East, West, North, South. These were picked up by Homer and appear in the *Odyssey* and *The Iliad* as Boreas, Eurus, Notos, Zephyrus.

Aristotle later expanded on this set of four winds to suggest that there may be ten winds, eight of which are principal, two of which were minor, but he struggled with the problem that some of the winds didn't have contraries. Following Aristotle, Timosthenes of Rhodes developed a twelve-point wind rose, adding even more layers to the many attempts to make sense of the various breezes and gusts that would come from different directions.

The study of wind actually gets quite a bit more complicated after that, as the Roman thinker Vitruvius (c. 80–15 BC) developed a twenty-four-point wind rose, Seneca (54 BC–39 AD) knocked it back down to twelve points, and Pliny the Elder (23–79 AD) suggests that modern thinkers only seriously considered eight. Says Pliny:

> The ancients reckoned only four winds (nor indeed does Homer mention more) corresponding to the four parts of the world; a very poor reason, as we now consider it. The next generation added eight others, but this was too refined and minute a division; the moderns have taken a middle course, and, out of this great number, have added four to the original set. There are, therefore, two in each of the four quarters of the heavens. From the equinoctial rising of the sun proceeds Subsolanus, and, from his brumal rising, Vulturnus; the former is named by the Greeks Apeliotes, the latter Eurus. From the south we have Auster, and from the brumal setting of the sun, Africus; these were named Notos and Libs. From the equinoctial setting proceeds Favonius, and from the solstitial setting, Corus; these were named Zephyrus and Argestes. From the seven stars comes Septemtrio, between which and the solstitial rising we have Aquilo,

named Aparctias and Boreas. By a more minute subdivision we interpose four others, Thrascias, between Septemtrio and the solstitial setting; Cæcias, between Aquilo and the equinoctial rising; and Phœnices, between the brumal rising and the south. And also, at an equal distance from the south and the winter setting, between Libs and Notos, and compounded of the two, is Libonotos. Nor is this all. For some persons have added a wind, which they have named Meses, between Boreas and Cæcias, and one between Eurus and Notos, named Euronotus.[13]

And then pretty much everyone in western Europe went into a deep intellectual slumber. The Dark Ages ensued. (I mean, sure, there was still a lot of stuff happening during the Dark Ages, but to stick with our cartoonish intellectual history, let's stay the course.)

*****

On the other side of the world, Islamic scholars such as Al-Jahiz (الجاحظ), the "bug-eyed" (776–869 AD) essentially laid the groundwork for what would become the modern science of ecology by conjecturing that the movement of animals closely corresponded to the weather.[14] Abu Yusuf Ya'qub ibn Ishaq Al-Kindi (أبو يوسف يعقوب بن إسحاق الصبّاح الكندي) – the "father of Arab philosophy" (801–873) – helped bring Aristotelian and Ptolemaic insights to the Arab peninsula, unwittingly preserving them for rediscovery hundreds of years later during the Renaissance. In his works he continues the tradition of insisting that weather is caused by celestial bodies.[15] Scholars such as Abu Bakr al-Razi

(أبو بكر محمد بن زكرياء الرازي), philosopher and alchemist, (860–925) carried on in the environmental determinist tradition of Hippocrates, turning his eye toward pollution in the air, but essentially going on to challenge the Aristotelian order of things. Ibn al-Haytham (أبو علي الحسن بن الحسن بن الهيثم, Latinized as Alhazen, 965–1040) wrote extensively about the refraction of sunlight through the atmosphere. Ibn Rushd (ابن رشد, Latinized as Averroes, 1126–1198), likewise, wrote extensively about how the universe – the sun, the moon, the Earth, and the weather – is perfectly tuned to permit human civilization to flourish.

More sophisticated instrumentation also came on the scene. The Islamic astrolabes – which are essentially metal star charts depicting celestial bodies – were used not just to locate one's position on the planet (say, when at sea) but also to anticipate the weather. Figure 1 is actually from Geoffrey Chaucer's "Treatise on the Astrolabe" (1391), which is one of the earliest English manuscripts – an instruction manual, effectively – to discuss the uses for a scientific instrument.[16] It is hardly Chaucer's most important work, but the term "Renaissance man" exists for a reason. So many of the well-known polymaths that you will recognize for their works in other areas were also extraordinarily involved in laying the groundwork for what has become the meteorological sciences.

Renaissance and Enlightenment Meteorology

All manner of scholars and academics were involved in this business of trying to understand the weather, but weather

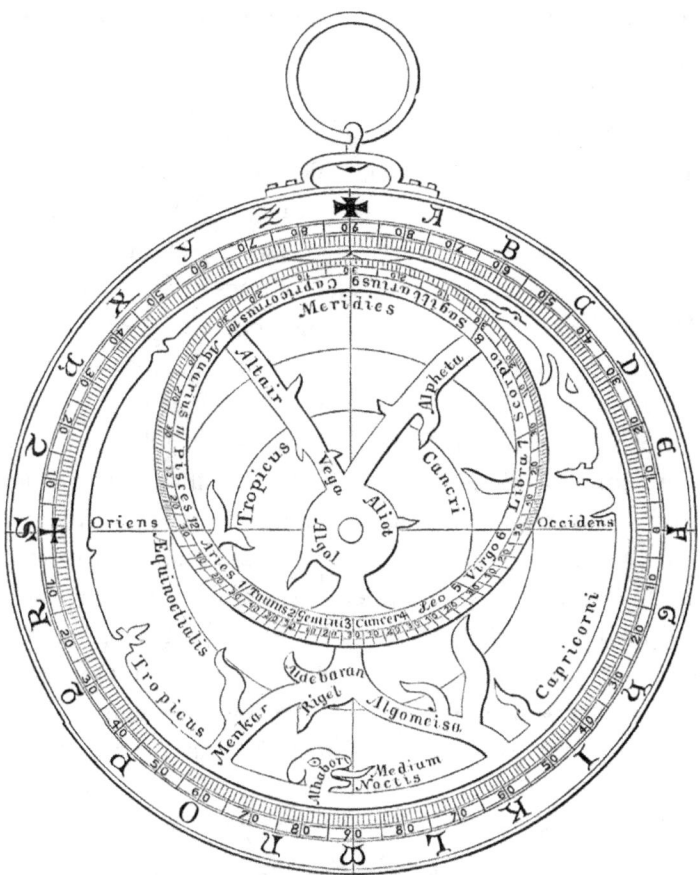

Figure 1 Image of an instrument called an "astrolabe" as described by Chaucer from the nineteenth-century edition of Geoffrey Chaucer's 1391 work "Treatise on the Astrolabe."

science in Western civilization really took off in a different direction once the cowl of the Dark Ages began to lift. The Renaissance was a rebirth, a rediscovery of lost knowledge, and the intellectual ideal at the time was to immerse oneself in everything.

Francis Bacon (1561–1626) is perhaps best known for his *Novum Organum* (A New Method), in which he lays the groundwork for the scientific method, but in a more obscure segment of that book, "Preparative toward Natural and Experimental History," he also paves the way for covering the history of such things as "Lightnings, Thunderbolts, Thunders, Coruscations," "Winds and Sudden Blasts and Undulations of Air," "Rainbows," "Blue Expanse, of Twilight, of Mock-Suns, Mock-Moons, Haloes, various colors of the Sun," "of Air as a whole, or in the Configuration of the World," and more than a few other strange collections of weather events.[17]

Galileo of Galilei (1564–1632), best known as a valiant defender of Copernicus's heliocentrism, also invented one of the first thermometers – the thermoscope – which functioned by using the heat of one's hands to warm a bulb that was then inverted and tipped into a small vessel of water; then, depending on temperature in the room, the water would rise to a height that varied by temperature. Galileo is also credited with naming the aurora borealis (meaning "northern dawn"). Among the many things for which Johannes Kepler (1571–1630) is known, he also wrote an obscure piece in 1611 – *A New Year's Gift: On the Six Cornered Snowflake* – which he positioned as a present for his friend Johannes Matthaeus Wacker von Wackenfels.[18] Essentially an ode to the snowflake, he outlines the reason for the "six-angle shape of the snow crystals." Kepler's gift is thought to be one of the first mathematical explorations of the symmetry of snow crystals, credited with kicking off the scientific subfield of crystallography.

Many philosophers know René Descartes' *Discourse on Method* as one of the most important and influential works of the past 500 years. This is the book in which Descartes (1596–1650) addresses the problem of skepticism and spells out most clearly his unique brand of "methodological skepticism," the ultimate claim of which results in his foundation-building conclusion that if there is one thing that he knows clearly and distinctly, is that he a thing which thinks. "I think, therefore I am." *Cogito, ergo sum*. What they may not know, or may have read and then forgotten – I certainly did – is that the full title of the book in which the *Discourse on Method* appears is *Discourse on Method, Optics, Geometry, and Meteorology*.

> Although the clouds are hardly any higher than the summits of some mountains ... nevertheless, because we must turn our eyes toward the sky to look at them, we fancy them to be so high that poets and painters even fashion them into God's throne and picture Him here using His hands to open and close the doors of the winds to sprinkle dew upon the flowers, and to hurl the lightning against the rocks. This leads me to the hope that if I here explain the nature of clouds in such a way that we will no longer have occasion to wonder at anything that descends from them, we will easily believe that it is similarly possible to find causes of everything that is most admirable above the Earth.[19]

Descartes, an unrepentant rationalist, thought that he could derive the basis of his meteorology through the contents of his discourse.[20] Though there were many misses, he was able to analogize wool to air, and using his somewhat confusing theory of vortices, he surmised that air has pressure that can be measured. As a consequence, some historians credit him with having developed the first barometer before Evangelista Torricelli (1608–1647).[21] (Most also know that he developed the Cartesian coordinate system, which has wide-reaching utility across the sciences, not just meteorology, so given the argument I'm making here, it would be a mistake not to mention that he also had an indirect hand in the mathematization of meteorology.)

The Enlightenment's repositioning to conceptualize weather in more natural, rather than supernatural, terms – or, differently put, the Enlightenment's *naturalization* of weather – brought with it innumerable benefits for humanity and civilization. For one thing, it stripped weather events of their purposive import, lifting an otherwise heavy burden from our mortal shoulders. The thing about gods is that they act with intention. They are deliberate in what they do; they do things for a reason, with a purpose, to be retributive, to exact justice, or to set the universe straight. By shifting to think of weather as a natural phenomenon, weather events were shorn of these intentional descriptions. According to the naturalistic worldview, violent hurricanes and droughts ceased to be punishments for the unsavory transgressions of civilization: the wrath of the gods meted out on their mortal subjects. The metaphysical nature of weather was being removed and replaced with a characterization more resembling the crashing of billiard balls.

The Enlightenment philosopher John Locke is reputed to have kept a weather diary for thirty-five years, from June 1666 until May 1703, in which he took at least two readings every day from a thermometer, barometer, and wind gauge.[22] The philosopher and mathematician Blaise Pascal (1623–1662) carried a barometer with him and showed that atmospheric pressure varies with changes in altitude, eventually leading to the convention nowadays of referring to atmospheric pressure in pascals, which is defined as one newton per square meter. Which raises the obvious point that a "newton" is, of course, named after Sir Isaac Newton (1642–1726), who didn't come on the intellectual scene until just after Pascal, but who has so many credits to his name as to be almost synonymous with the idea of the Renaissance man. Among his many accomplishments, Newton is widely understood to have developed calculus slightly before the philosopher Gottfried Wilhelm Leibniz (1646–1716). But, as seems to have been the trend among Renaissance men, Leibniz is also said to have been instrumental in developing a nonliquid barometer. All of these mathematicians, I should further note, will become increasingly important as we delve deeper into the philosophy of weather prediction and the various attempts to manage risk through probabilistic tools.

Benjamin Franklin (1706–1790) is well known among American schoolchildren for having attached a key to a kite string, aiming to understand the relationship between lightning and electricity. But in his *Meteorological Imaginations and Conjectures*, Franklin speculated that there was a place just above the Earth where it was always winter, and that this could account for hail.[23] Also in that document he conjectures about a possible connection between volcanic activity and weather, and specifically raises the possibility that the 1783 eruption of the Icelandic volcano Laki was responsible for wet weather and icy conditions. Of course, it was also about this time that Antoine Lavoisier (1743–1794) was in the process of discovering oxygen (by way of hypothesis and reasoning), thereby debunking what was then the prevalent view of the composition of the air: that the world was made primarily of phlogiston.[24] (You may remember this historical inflection point if you've ever picked up a text on the philosophy of science or studied the work of Thomas Kuhn.)[25]

Franklin is not alone among pioneering Americans. Several of the US Founding Fathers were likewise dabbling in weather observation. Thomas Jefferson (1743–1826) and James Madison (1751–1836) both set to the task in the early years of the United States of America of documenting weather in the newly established country. We know, for instance, from Jefferson's own records that on July 4, 1776 – the very first day of the new Republic – it was 76°F in Philadelphia. Jefferson was, as it also happens, something of a climate determinist who thought that tropical climates gave rise to laziness, promiscuity, and general moral turpitude.[26] His efforts were partly responsible for arguing that the swampy shores of the New World could and should be colonized by the British.[27]

And that's it. That's where the wave of familiar Renaissance men stops cold and we enter into a new phase of intellectual inquiry that involves far less well-known men. (No, I'm just kidding. There were tons of other notable figures and early scientists, driven by the ideals of the Enlightenment, who also contributed to the foundations of weather research: Giovanni Cassini, Robert Boyle, Christian Huygens, Robert Hooke, Edmund Halley, Vitus Bering, Daniel Fahrenheit, Anders Celsius, Denis Diderot, Henry Cavendish, among many, many others.)[28]

The history of modern meteorology is built upon a foundation of centuries of intellectual labor that involved moving beyond the various presuppositions and eventually reorienting around the objective not just of understanding what was going on but truly making sense of the machinery that makes weather work.

Despite all that was known and learned, weather remained a fairly elusive phenomenon until relatively recently. Throughout the Enlightenment, just as in past periods, weather forecasts were captured in catchy aphorisms, proverbs, and dubious prognosticating devices like the *Farmer's Almanac*, dating back, as we said at the beginning, at least as far as Hesiod. Most people know the phrase "red sky at night, sailors delight. Red sky in morning, sailors take warning" – which allegedly dates back as far as biblical times[29] – but there were plenty of other common phrases as well, many of which we passed down as helpful predictive advice not much more sophisticated than the sayings that Theophrastus had offered 2000 years prior:

- "The higher the clouds, the finer the weather."
- "Clear moon, frost soon."

- "When clouds appear like towers, the earth is refreshed by frequent showers."
- "Rainbow in the morning gives you fair warning."
- "Ring around the moon? Rain soon."

For reasons that we've already talked about extensively, those who are dependent on nature still need to know how to predict what's going to happen. Farmers, of course, but also sailors, fishermen, builders, merchants, and militaries are deeply dependent upon weather systems. It took centuries before all this collected know-how coalesced to form the scientific enterprise we now know as meteorology.

Modern Meteorology

Some attribute the modern science of weather forecasting to Robert FitzRoy (1805–1865) – a vice-admiral in Britain's Royal Navy, the erstwhile captain of Charles Darwin's HMS *Beagle*, a two-year governor of New Zealand, and the founder of the UK's national weather service (the Met Office) – who coined the term "forecast" to describe the science of anticipating the weather, which was critically important to him as a mariner. (Darwin, notably, was shaking up presuppositions about fate and accident in another realm, essentially suggesting that biological adaptation occurs by dint of luck, by accident, as it were; the two would both go on to be controversial figures in their own fields. I highlight this here because it is important both as an historical side note and as a critical reflection of a trend that was happening all across the sciences.)

But there were obviously more than a few others who were also working to make sense of weather using the tools of the sciences. I won't bore you with a long list of figures who have contributed to the development of our thinking on meteorological matters. Suffice it to say that this list is long – extremely long – but what is distinctive is that it is also characterized as a time when there was considerable diversification and specialization of disciplinary focus. That is, the much-vaunted Renaissance men – jacks of all trades, masters of none – mostly disappeared from the scene and were replaced by specialists.

The British Meteorological Society, later the Royal Meteorological Society, was formed in 1850, around the time that FitzRoy was leaving active duty. As was common at the time, the task of knowledge production was one of the society's primary objectives, mostly so that other educated observers could learn of discoveries and discuss ideas. To advance this effort, the members formed an editorial board and launched a journal. *The Quarterly Journal of the Royal Meteorological Society* dates back to 1873 and has been publishing about eight issues annually for the past 150 years. A similar society and journal sprung up in North America shortly after that. The *Bulletin of the American Meteorological Society* is considerably newer and publishes twelve issues per year. It was started in 1920 and continues to this day, more than 100 years later. There is a ton of material contained in these two collections of volumes, covering a wide range of weather-related topics. I read through many tables of contents, titles, and abstracts from the early years so you don't have to, but I will say that these journals provide a fascinating window into the world of meteorology that I had only scarcely known existed before I undertook this project.

The first few issues of the *Bulletin*, for instance, are about the establishment of the American Meteorological Society: membership, bylaws, mission, purpose – that sort of thing. It takes a few issues for the scholarly publications to get rolling. But they begin where you might think they should begin, with short notes about meteorological instruments, thermograms, "sun kinks," the errors of household thermometers, and even the allegation that New York City architects were conspiring to keep America ignorant by refusing to place weather vanes on skyscrapers.[30] By July of its first year the American Meteorological Society was even publishing issues raising questions about whether climates *can* change, an issue that most of us take for granted now but that in 1920 seemed an absurd proposition.

To quote the author of this particular note:

> A few years ago a professor in a well-known college expressed a belief that the winter temperature was becoming higher at the rate of about one degree a century. Assuming this to be true, the temperature of the Mediterranean basin in the time of Theophrastus must have been about that of the Arctic coast of Siberia to-day! *Quo quid absurdius!*[31]

As meteorology has become more professionalized, it has also become much more rigorous and exacting, and the latter half of the twentieth century bore witness to this transition.[32]

The past seventy years of meteorology have seen an explosion in article production, data collection, and instrumentation innovation. Up until relatively recently, it was only possible to measure local weather conditions manually with legions of monitors and observers, and then to aggregate this information into a ledger. But now, our satellites enable us to monitor our global climate in a way never before thought possible. (It is arguable, but I think true, that the very idea of a "global climate" as currently understood – that is, as a longitudinal, global temperature average – is in part an outgrowth of this particular kind of practice.)

We have mapped the globe at each of the various atmospheric levels and can now model weather patterns months into the future. We can anticipate global climate change by applying the principles of fluid dynamics. Whereas before this had to be done using a slide rule, our computing power enables us to run through many hundreds of model futures. We can compare these models against different models. We can anticipate earthquakes by understanding tectonic movement. We also have the techniques of the climate sciences, including general circulation models, parameterization, and ensemble methods, so we are able to approximate with some accuracy changes in microclimates that we were never able to measure directly.

Since my objective here is to offer only a cartoonish intellectual history of the main ideas in meteorology, rather than a proper historical account of the various figures and milestones in its development, I won't bore you with the various geopolitical and economic circumstances that gave rise to modern meteorology. It is only important to note that this transition reflected trends across the intellectual spectrum that were simultaneously being criticized by other philosophers and social critics.

Building the Conceptual Apparatus

It would be easy to think that our ability to predict the weather has developed mostly because we have much better science. But what does that really mean? Author Tristan Gooley suggests that four factors have changed the accuracy of our forecasting science: (1) our understanding of underlying meteorological processes, (2) the abundance of data, (3) much more powerful computing, and (4) interconnectedness.[33] To an extent, I think he's correct. Having a more developed scientific theory, better data, powerful computers, and a global network of informed practitioners helps immensely in promoting the aims of the sciences. But embedded in the scientific process is a more subtle set of considerations.

I routinely try to make this point when I teach my graduate-level Conceptual Foundations of Environmental Studies class: that science is not a simple project of data collection and observation, though we often talk like it is. There's a whole apparatus of conceptual work and prior knowledge that undergirds scientific inquiry. Scientists make *observations*, to be sure, but before they do even that, they've already done a lot of the thinking that builds up to those observations. What I mean is that they first have to *conceptualize*, or come up with the ideas for, whatever it is they're exploring. Once they've done that, they also have to *theorize* the mechanisms that explain their observations, and they have to pull together models that substantiate or anticipate those models. In the case of weather, this means not just making observations but also understanding the processes, bringing all that data into one place in a timely fashion, and crunching that data by plugging it into the model. In other words, the three legs of the contemporary scientific theodolite are *conceptualization, theorization,* and *observation.* Concept, theory, and data work in conjunction to bring us scientific clarity on complicated problems like weather.

But of course, even with all of these very smart people working over the millennia to conceptualize, theorize, and observe the weather, meteorological prediction has remained an inexact and ridiculed science. And this, presumably, is because the science of weather prediction is "extremely complicated."

But what even does that mean?

On one way of thinking about meteorological complexity, it's all just a matter of figuring out the various different knobs and levers that make the weather work. Not only is it the case that meteorologists have to know things about the way in which the clouds and the winds are moving, about high- and low-pressure systems and so on, they also have to know about the landscape itself. Forested areas, mountain ranges, large bodies of water, cityscapes, roadways: these all affect how different conditions, sometimes even the same conditions, will play out. There are still a lot of unknowns. It would seem, from casual observation, that these unknowns explain the *uncertainties* of weather forecasting.

On another way of thinking about this complexity, there's considerable *stochasticity* (or randomness) associated with weather systems. Drop some steel marbles onto a carpet and you will see them settle into position relatively quickly. Drop them instead on a smooth slate surface and they may ping around for some time before rolling to a long, slow stop. Drop those same marbles into a pachinko machine, and you will see them pitter-patter through the pins with stochastic randomness that even the world's fastest supercomputers can only barely attempt to approximate.

But I think that some of the problem with anticipating the weather is made that much more complicated because these systems are more *indeterminate* than uncertain or stochastic.[34]

I know I said I wasn't going to focus on climate change in this book, but I think it will help illustrate the difficulty of anticipating weather. In the age of climate change, longer-term predictions of the weather are that much more challenging. Where before the term "climate" had served as a nebulous state of background conditions used loosely to inform us about impending events, now it is no longer quite as reliable. As new climate regimes begin to form, we face what ecologists sometimes call "no analog futures." This means that given the fluid nature of change, we can never be in a position to anticipate what kind of weather events might affect what areas, where, and how. That is, we can't easily analogize how various weather systems will respond to prompts and drivers because the prompts and drivers are mixing in ways that they haven't yet mixed before.

Areas that otherwise would not commonly flood now may be subject to flooding. Areas that otherwise might not face drought will suddenly experience unusually dry conditions. We do not know, in part, how individual storms or events will affect the changing landscape in those areas. We have to extrapolate from other parts of the world, recognizing that soft soil that would normally be in a position to take up more water may not do so if it has been dry for an extended period.

This is complicated in part because, over the longer term, our present-day observations about the way in which weather works affect the kinds of decisions that people make about where they're going to build, what kind of lifestyle they're going to lead, what sorts of jobs they should pursue. Not only this, but they do so strategically, which is to say that they make investments based on past meteorological observations about weather patterns. This is, again, more an issue for climate prediction than for weather prediction, but the point bleeds over into the meteorological sciences.

Industries like agriculture may choose to open a new area for development on the basis of anticipated rainfall, only to discover that rainfall is not as consistent in that area as it once was. Other industries, like outdoor recreation, may choose to develop an area because it is expected to receive a lot of snow, only to discover that the snow and precipitation doesn't materialize as expected. Still yet other industries, like tourism, may choose to develop a coastal region based on historical information about flood risk, and may even have insured themselves against flood damage using the same meteorological spreadsheets that the insurance companies and the reinsurance companies use, only to discover that the entire framework of suppositions that has informed their decision is no longer in play.

This may seem similar to an observation we've already made, because of course it's true that people and private actors will continue to act in a way that makes the most sense for them (e.g., is economically rational or self-interested), but the point here runs deeper than that. People act on the basis of *information* that was generated by the meteorological literature itself. In other words, the very same information that was supposed to help us get a grip on what the future holds creates the conditions that make it possible for others to respond to reports of the future, thereby obviating the critical insights of that literature in the first place. The climate futures made real by climate change are not simply *uncertain*, they are *indeterminate*. What I mean here is that future weather and its effects are not just unknown or closed off to even skilled observers (and therefore *uncertain*), but that it is not yet determined, that it will unfold in a way that depends on factors that themselves change with the prediction (so it is also *indeterminate*).

This indeterminacy is a *real* and *serious* problem for those who are in the business of predicting the weather and anticipating its impacts. The impacts of storms and weather events – hurricanes, tornadoes, floods, fires, droughts – will very much depend on how people respond to weather predictions. If a weather forecaster anticipates the "storm of the century" and people heed this warning, the outcomes of the storm of the century may well be less significant than another storm that is not anticipated to be the storm of the century. This is not a big deal, of course, because it is generally better that people heed the warnings of meteorologists, but it does call into question the extent to which the science of "prediction" does the right kind of predicting. In retrospect, the meteorological anticipation of the "storm of the century" will be *wrong*, because people will have made it wrong. Alternatively, meteorological predictions of lesser weather events (to which people respond less urgently) may then also be wrong, because the affected people will not have taken steps to protect themselves and the outcomes will be worse.

From Ideas to Words

It is an oft-uttered cliché that Eskimos have fifty different words for snow, each describing a slightly different state of snowfall.[35] When I was younger, I used to find this claim about the Eskimo language fascinating. Where I grew up in Virginia – a mid-Atlantic state on the coast of the US – we commonly didn't need to distinguish between the different kinds of snow. We basically only had snow or rain, with intermediate forms like sleet or freezing rain. It fascinated me, first, that there could be such variability of snow that I otherwise wasn't aware of. But also, second, it fascinated me that language might be the kind of thing that was picking out states of the world that otherwise go unnoticed.

> **Box 8 Words for Snow**
>
> Singer/songwriter Kate Bush has a whole album (and a song) with the title "50 Words for Snow", though her list is a litany of playful neologisms – faloop'njoompoola, whippoccino, shimmerglisten, sorbetdeluge, anechoic, etc. – that don't clearly bear on any Eskimo heritage.[36]

It is easy to believe that language picks out real features of the world: that when I say "ball" I mean that spherical thing in the corner, or when I say "dog" I am referring to a type of animal that naturally exists in a set of animals distinct from "cats." It is easy to think, in other words, that language "carves nature at its joints," that the world comes to us pre-divided, that wherever there is a natural delineation between one part of the world and another part of the world, we can slap a term on that different part and add it to the dictionary.[37] It is further easy to believe, in virtue of this, that different languages should be isomorphic with one another, such that the word for "snow" in English maps directly onto the word "Schnee" in German or the word "снег" in Russian. The thought that Eskimos have over fifty words for snow, where in English we only have one word for snow, blew my mind.

Unfortunately, this is a widespread myth that has little basis in fact. Indeed, the myth is so pervasive that journalists have coined the term "snowclone" to refer to the phenomenon of lazily repeating a meme or a cliché without investigating its origin.[38]

When you dig a little deeper, it starts to become apparent that the multiple words for snow are words that are not, in fact, so divergent from words that we ourselves might have. Qanuk is Eskimo for "snowflake", Kaneq is "frost", Kanevvluk is "fine snow", Qanikcaq is "snow on the ground", and muruaneq is "soft deep snow."[39] It's easy to see how someone living in a snow-rich environment would need these words, and need to distinguish them from one another.

Living in Colorado, it is further clear that communities much closer to home do in fact also need different words to refer to snow. The skiers here distinguish between powder days and heavy snow days, between flurries, freshies, glop, groupel, scrape, powder, crud, cream, chunder, dust on a crust, mashed potatoes, snirt (snow mixed with dirt), and corduroy. I'm not cool enough to know the rest of the vernacular, but you get the idea. These are linguistic concepts that are born of necessity, which we use to describe the weather (or other phenomena) when it is relevant and meaningful to us. Snowboarders, like Eskimos, need multiple terms to describe minor differences in the snow because it matters to them. Words are useful. Farmers, no doubt, also need different descriptors for different kinds of weather. Maybe a light rain, or a gentle mist, or a gathering fog.

There's a fabulous episode of the *Radiolab* podcast in which hosts Jad Abumrad and Robert Krulwich discuss the phenomenon of color.[40] Among several topics, they suggest that throughout many of the known ancient Greek writings, there is no reference to the color blue. Their audacious conclusion, which is somewhat unbelievable, is not that the color blue was neglected by authors and poets but that the color blue *didn't exist*. What could they possibly mean by this? Casual observers of the sky will insist that it is obviously the case that the sky has always been more or less the same color. They may add that it is also the case that the color blue as a frequency of light has existed. But what Abumrad and Krulwich suggest instead is that people of antiquity hadn't yet identified the color blue *as* a color and so didn't speak of it or come up with a word to describe it. Without a word to describe it as a color, it was, in a way, an undifferentiated and invisible background feature of the world.

The same sort of observation might be made of weather, where weather in its various forms has always been around but often as a kind of as unobserved noise or an invisible background. Though people were no doubt dependent upon sun or rain, aware obviously that there were clouds and fog and wind, weather was a much more abstract notion that took intellectual reframing to become a discernible thing.

Likewise, Sir Isaac Newton is said to have discovered and measured humidity, giving a name to a phenomenon that otherwise went unexamined. The discovery and identification of humidity was no doubt a great advance in the history of weather prediction. So too with barometric pressure, radiation, ozone, "dephlogisticated air," cloud layers, radar, the jet stream, the magnetosphere, emissions, and almost any weather phenomenon you care to investigate. This is exactly what happened as meteorology developed: We created a language to describe weather phenomena that we otherwise might have missed. Part of the scientific endeavor involves precisely this: identifying features of the world, giving them a name, and then using this name as a basis for making further observations. Just as with many words for "snow," it is easy to believe that the world that the philosophers of the Enlightenment were encountering was the same world that the philosophers of antiquity were encountering, and as such that the language from yesteryear should be isomorphic with the language of the 1700s.

I make this point about language because it's an important part of scientific discovery that we have, over the years, developed terminology to help us understand the world around us. It doesn't just help us make sense of the different kinds of weather that it describes but also indicates to us how we might better anticipate and prepare for that weather. The better we are at using language to identify aspects of weather, the better we can be about understanding it. Just as a child might look at a bird and see only a creature with eyes, a beak, and wings, whereas a trained ornithologist might look at a bird and see eyes, beak, breast, wings, tail, legs, scapulars, coverts, and rectrices, so too can the contemporary meteorologist look at weather and see much more than you or I.

From Words to Numbers

But the creation and development of terminology to describe different states of weather is just a small part of a much bigger picture. For my money, the bigger innovation in the development of meteorology has been the advent of *statistics* and, more directly, its role in *risk assessment*, which we discussed briefly in Chapter 3. Statistics has wildly transformed not just the way in which we predict and anticipate weather but also how we think about ourselves and the actions we take in relation to weather. Over the course of the past few hundred years, we have developed different metrics to capture what we take to be elements of weather: barometric pressure, humidity, temperature, wind speed, etc. By measuring the universe in spatial increments, partitioning temperature on a scale, and breaking up time on a timeline, we created for ourselves the possibility of understanding the world in a way that allows us to control it. Carving up "weather" into small, cognizable bits gives us something to hang on to; we defang and demystify the chaotic nature of the force that we have been exploring and transform it into a list of facts and numbers.

Consider how unusual this is. Temperature doesn't have to be understood in these discrete, epistemic chunks. Young children certainly don't think about temperature in degrees. Most of the way in which we experience temperature is rather subjective, as the feeling of hot or cold, not unlike the Tucson weatherman's characterization of the weather as "Hot." We dip our foot into the ocean to test whether it is too chilly to swim, and from there we make a judgment about its temperature. No doubt we've all had the experience of doing precisely this, judging the ocean to be too cold, forcing ourselves to dive in, and then adjusting our assessment of the temperature once our body acclimates to the water. Objective measures such as Fahrenheit, Celsius, and Kelvin change our relationship to temperature itself.[41] We cease to understand it as a matter of hot and cold and only think about hot and cold as existing within ranges on the temperature scale.

How many times have we been sitting around the kitchen with our grandparents as they grouse about how cold or hot it is in our dwelling, only to seek an independent, objective metric that will settle the matter once and for all. "Don't be crazy grandma, it's 68 degrees in here!" all while she swaddles herself in her sweater. Of course, when Protagoras (490–420 BC) observed that "man is the measure of all things," he was himself raising a question about the extent to which truth is filtered through human experience.

Maybe a more natural way of thinking about this idea is to consider a pain scale. Physicians and emergency room technicians aim at much the same idea when they try to get us to articulate our pains. They may present us with a ten-point scale of happy or sad faces and ask us to compare our internal subjective states to these happy and unhappy cartoons. Measuring wind gusts or temperature works something like this, putting an objective gloss on an otherwise subjective phenomenon.

The point really gets lost if we think about this only in terms of temperature. This kind of "carving up" and "precisificiation" happens throughout meteorology and the sciences in general.

One of the other ways in which weather forecasting has become more precise has been through the identification and attribution of probabilities to otherwise seemingly random weather events. At first these were fairly rudimentary probabilities, likely rough guesstimates of the likelihood of some event coming to pass. But the mathematical tools of statistics have always been in the background of some modern approaches to the sciences, and this is just as true in meteorology as elsewhere.

More recent innovations in weather forecasting have leaned heavily on computer modeling and the massive explosion in inexpensive computing power. Rather than manually grinding through charts and maps, anticipating the movement of high- and low-pressure systems, computers can crunch through fine-grained data to assess an ensemble of possible outcomes. These so-called ensemble methods anticipate multiple possible states of a weather system and from there produce a range of conceivable forecasts. On the basis of these ensemble of forecasts, meteorologists can then calculate the probabilities of any given forecast occurring, reframing hurricane paths in terms of probabilities. As our analytical tools have gotten more sophisticated, and our processing power has expanded, our ability to forecast the weather has grown equally impressive.

I mentioned earlier that those mathematicians of the Enlightenment would become increasingly important as we move forward, and I return to them here. Mathematicians like Blaise Pascal, Thomas Bayes, Pierre-Simon Laplace, Carl Friedrich Gauss, and even Charles Sanders Peirce developed over years increasingly more sophisticated probability theory, which helped transform the project of weather prediction from an art into a science. In some respects, probability is nothing new. As a rough tool for thinking about things it dates as least as far back as gambling; and indeed, gambling has existed since the beginning of recorded time. The Greek gods Zeus, Poseidon, and Hades were said to have cast dice to determine where they should rule. But the abstract ideas of probability theory didn't emerge in full until Enlightenment mathematicians like Pascal and eventually Bayes put pen to paper and harnessed the tools of mathematics to better describe the modalities of the future.

> **Box 9 Pot of Gold**
>
> Renowned seventeenth-century rationalist and pantheist Baruch Spinoza was addicted to games of chance, which seems in part to have given rise to his theory of rainbows, resulting in two seemingly unrelated treatises: "Algebraic Calculation of the Rainbow" and "Calculation of Chances."[42]

This itself is a curiosity, in a way, because one would think that probability wouldn't be that difficult to discover. But in Peter Bernstein's wonderful book *Against the Gods: The Remarkable Story of Risk*, Bernstein outlines the way in which critical innovations in language changed our thinking about numbers and in so doing made a more calculative mathematics possible.[43] Certainly the ancient Greeks thought that the world could be understood mathematically – and Plato most notably makes this case to Phaedo – but mathematics for Plato was said to reside in the realm of ideas, recollections of concepts that we all already know and, given enough reflection, can recall from deep within our memories. At least some of Bernstein's argument rests on the detail that the numerical systems used throughout antiquity were mostly deployed in the service of tabulation, which made abstract calculation extremely difficult. Additionally, many systems didn't have the number, or the idea of, zero (which instead emerged out of Hindu-Arabic numerical systems) which made a whole numerical system visible. Bernstein quotes philosopher Alfred North Whitehead on this point:

> The point about zero is that we do not need to use it in the operations of daily life. No one goes out to buy zero fish. It is in a way the most civilized of all the cardinals, and its use is only forced on us by the needs of cultivated modes of thought.[44]

This has had downstream phenomenological effects as well. Forecasting science has changed how we think about and experience weather. Where historically we might only have thought of the weather as a string of events with immediacy – it was warm this morning, but now it is snowing – the very presence of weather maps and forecasts changes the momentariness of these moments. They become states of affairs to anticipate, to look forward to.

Not only do they shift our thinking about the weather we are experiencing and anticipating right here, right now, but weather science and weather mapping provides us with access to and awareness of information about weather events across the planet. We can see that there is flooding in Pakistan or that there is a heat wave in Europe – and we may from this news gather considerable confidence that weather is not isolated by the immediacy of the conditions around us and is better understood as a holistic, all-encompassing phenomenon.

It is arguable, but I think true, that the idea of climate as conventionally construed is itself a manifestation of this broad meteorological and statistical mapping. Medieval farmers would never know what the weather was like even a few hundred kilometers away. They would instead be left to surmise from agricultural or landscape clues about broader seasonal trends. They might be able to talk intelligently about the seasons – with regard to rainfall or freezes or frosts; they might look at the flora of an area and ascertain the extent to which a landscape was more desert than tundra; but they would have had difficulty characterizing climate in mathematical terms, as a longitudinal approximation of averaged weather phenomena reflecting conditions and patterns of temperature, humidity, pressure, wind, precipitation, and other atmospheric and oceanographic conditions.

In a way, the influence of these conventions is even evident in our own contemporary vernacular. Apart from a college classroom, consider how you think about climate, not as longitudinal averages and mathematically uniform patterns, but instead as vague cartoons of regional weather types specific to different biomes (desert, jungle, tundra, forest, coastal, permafrost, arctic, alpine, etc.).

All of these statistical tools play a part in how we approach and think about weather, much like the technologies that we covered in Chapter 3. They change our relationship to the risks that weather presents to us. Where before we established that weather was always around us and forever changing, almost entirely out of our control, here we must acknowledge that it follows patterns with lawlike consistency. Within reason it is true that we can anticipate some variability in weather, but the creation of statistical tools gives us reason to believe that we can now understand it and anticipate it with increasing precision. The quantification of weather into discrete units has given us the tools to anticipate how those units act in the world. The reification of weather as a chain of discrete weather events – rain, snow, sleet, hail, tornadoes, hurricanes – lends itself to the mistaken impression that weather is not an abstract force. And as we shall see in Chapter 5, this all lends itself even more to the mistaken impression that we can control nature, that we remain dominant over it.

The basic idea builds on a point that I have been making about the risk society: that as we have built up our world to address the various challenges that weather and nature more broadly throw at us, we develop technologies to mitigate these challenges, but in so doing we also create for ourselves new dependencies and introduce new hazards. When we build dams to block rivers and prevent them from flooding our cities, this is a great advancement in our safety, but the building of those dams comes at a cost. We must forever live in the shadow of the risk that the dam will fail. In this way, our risk is reflexive, meaning that it turns back in on itself. We take risk mitigation steps to avert hazards, possibly by thinking of those hazards in terms of the probabilities, and those risk mitigation technologies themselves create new risks. Dispositionally this means that our attitudes about the world around us are forever shifting in order to respond to the newly constructed world that we have built and also in response to the idea that the weather is predictable

In turn, distancing ourselves from the weather itself, and breaking it down into discrete features that we can wrap our minds around, transforms our relation not just to the weather but also to others who might be in a less privileged position that enables them to process weather events in the same way that we do.

*****

In a way, weather has always been caught in the interstices between the supernatural and the natural, between the sky and the earth, between the gods and nature, between God and Man. As the world has become demystified, however, we have ceased to look at individual weather events as acts of God, and we've also learned to characterize them differently. Though a young child may still look up to the heavens and think of supernatural tears when it rains after a sad event, we in the present age know better than to associate precipitation with supernatural forces.

With this in our back pocket, we should once again return to our working definition and see if there is more to say. It is perhaps now important to stipulate that this force we have been talking about has features that are both chaotic and, at the same time, lawlike (or law governed). On the one hand, its chaotic nature seems always to remain out of reach of those that it affects. Much as we pursue an understanding of weather, and as advanced as weather science has become, there is still some sense in which its complexity continues to surprise us. On the other hand, weather abides by, or at least appears to abide by, the laws of nature.

> Weather is a *chaotic but lawlike* destructive and productive force that, coupled with human creativity, is an engine of change and innovation.

All of which raises the enticing prospect that if we can just wrap our minds around weather, if we can just excavate the hidden laws that govern weather, if we can just get the science right, then we pass this critical information off to our engineers and technologists, who then, with a touch of magic, might be able to make it do our bidding. From indoor air conditioning to outdoor cloud seeding, dreamers and stargazers have long wished for the day when they can call the proverbial shots.

5 Catching a Cloud
Controlling the Weather

If you've ever tried to grow a garden, then you will be familiar with agricultural worries about weather. No matter where you live, as a gardener you must be attuned to the climatological norms for your region. In Colorado we are told to wait until after Mother's Day to begin planting our seedlings, after which there is usually a low but non-negligible probability of a late frost. It's a little different in California and Florida, not to mention Arizona, Minnesota, or Alaska. Planting advice differs across the United Kingdom, Australia, Europe, Latin America, India, and pretty much wherever you go. Anywhere there is weather, there are concerns about how crops will grow. Gardeners worry about late frosts in spring, hailstorms as their seedlings begin to mature, blazing sun in the middle of summer, lack of rain during drought conditions. Fortunately, there are readily available regional maps that clarify planting times for each climate and zone, and over years of accumulated knowledge the gardening community has gotten pretty good about knowing when to put seeds in the ground.

The USDA began publishing a Plant Hardiness Zone Map in 1960 that amateur horticulturalists and farmers can use to determine what crops to grow and when to plant. (Pioneering botanist Alfred Rehder published his first climate-based "plant hardiness" zone map in 1927, which was later updated in 1938 by the horticulturalist Donald Wyman.) Since then, the map has been expanded to cover most other regions of the world. The USDA map divides the country into thirteen different zones that are calibrated to the diffuse microclimates of the United States.

Indeed, these climatological constraints fall neatly and tidily in the dispositional categories we discussed earlier: living with the weather. But over the centuries, and to a large extent much more recently, we have developed technologies that enable us to shed the constraints of geographical location by replicating weather and climate conditions more conducive to our objectives. It is now possible to grow oranges in Alaska, cabbage in Costa Rica, and mangoes in Denmark. Even very small countries that otherwise might be at a distinct resource disadvantage given location and size, like the Netherlands, have made great strides in these efforts. Now, somehow, the Netherlands has become the "second largest exporter of agricultural products by value."[1] They have gotten so good at the science of artificial agriculture, in other words, that they are beating mother nature at her own game. The good news is that much of the science that supports their agricultural success is widely available and fairly inexpensive.

At my home I have an elaborate hydroponic setup, and I control as much of the environment in which my plants are growing as I can. Given this control, I'm able to produce many more tomatoes and cucumbers than any of my neighbors. In the early life of my plants, I sow seeds in small cubes of rockwool (which is inert inorganic material spun into a filament not unlike asbestos), that will keep the seeds suspended and damp so that when they grow they have a medium in which to hold themselves by their roots. I place them under grow lights, which they use as their primary source of energy, converting artificial rays into sugars that they can then consume in order to grow. If the light is too far away from their base, then they will grow leggy, so I need to be sure to put it as close as possible to give them the most intense artificial sunlight they can receive. I keep them covered and protected in the stable environment of my garage, using small fans to replicate the gentle movement of the breeze, hardening their stalks so that when I move them outdoors they will be better able to tolerate the weather.

When Mother's Day comes, I move these vulnerable seedlings outside, but I shield them with shade cloth from too much sun, buttressing everything with mesh netting to protect them from the harsh Colorado winds that threaten to rip their delicate leaves from their stems. I fill two seventy-five-gallon reservoirs with water and infuse that water with nutrients. I run the nutrient solution from a reservoir on the ground, through PVC pipes and tubing, up to my plants. Like a drug pusher, I mainline nutrients right to their root systems, and they grow big and bushy and green, producing more fruit than you can believe.

The roots, soon much larger than the rock wool in which they were initially planted, spill out through their net cup, absorbing water and nutrient solution as if there were no threats at all in the world. If I cut off the nutrient solution, they will look sad and depressed and angry in a day. Within two days, they will die. In some respects they are too fragile not to receive their daily dose of juice. They don't know hardship. They don't know adversity. They're spoiled, addicted brats, but I love them for the fruit they give to me.

*****

Some of our first and most primitive forms of controlling the weather were of course crude attempts to capture precipitation and redirect it to do our bidding. I covered this briefly in Chapter 3. In an abstract sense, conduits, aqueducts, and acequias are a form of weather control, if we consider the harnessing of the weather to be both an effort to get weather to operate on our time schedule and an attempt to use its output to suit our objectives. Indeed, the metaphor of "harnessing" is precisely what we might do with horses to control them. But to categorize these kinds of efforts as weather control is to commit a similar error to that which we were making at the beginning of the book when exploring possible definitions of weather, confusing rain with water, confusing sunshine with light. Weather, we said there, is not a thing or a substance or a precipitate alone. It simply can't be. So harnessing the products of weather (precipitation) instead of the weather itself ought not to fall in the same category as controlling the weather.

Nevertheless, we can't really understand modern efforts to control the weather unless we return again to some of the ways in which we've sought to harness the weather, since those efforts essentially try to make it the case that not just the product, but the timing and intensity of the weather, operates according to our devising.

Indeed, we have engaged in primitive forms of weather capture and redirection since at least the Neolithic age. Now, of course, we are much more advanced. We have thousands of retention dams that store water for billions of people, redistributing it for irrigation, recreation, consumption, and industrial use. We have sprinklers that operate on timers connected to the internet, attuned to the weather-forecasting services that pump out information on a minute-by-minute basis. So too have we developed techniques to resist, regulate, and modify other aspects of weather. We shade our crops from the blazing midday sun, shield them from wind, build retention walls to protect from flooding, and so on.

But consider the expansiveness of our attempts to control the weather, and how these are far from limited to agriculture. We recreate snow on ski slopes to ensure that it comes at the right times and keeps our ski lifts open. Doing this doesn't involve trucking in snow that fell elsewhere, as if into a giant snow reservoir, but rather utilizing cold conditions and what we know about crystal formation to spray tiny droplets of water over ski slopes such that snow will form. This does start to look a little less like redirecting precipitation and more like controlling the weather. But I think we shouldn't be satisfied with that.

We have only relatively recently developed the possibility of capturing and harnessing other elements of the weather with the introduction of more sophisticated technologies. Grow lamps can now pump out light with enough intensity – about 30–50 watts per square foot – to provide plants with the light they need for photosynthesis. But artificially creating sunlight is a somewhat inefficient process, essentially converting sunlight into electricity to convert back into sunlight. At the same time the inefficiencies associated with capturing, converting, and recreating sunlight have the added benefit of avoiding the downsides of other elements of weather.

Many of our buildings nowadays are climate controlled, which means that we do what we can to create a stable interior environment that is conducive to human comfort. Like Goldilocks, we want to keep the weather not too hot, not too cold, but just right. For many people in the US and western Europe, this means setting the temperature somewhere in the region of 68–72°F (20–22°C), no matter whether it is 110°F (43°C) outside or below freezing. For my father, this means keeping the thermostat around 80°F (27°C), and though I try to reason with him about the environmental impact of his preferences, he nevertheless insists on being the god of his own domain.

From animal sacrifices to rain dances, from group prayer to cloud seeding, humans have tried for centuries to make the weather work in their favor. In this final chapter, I look briefly at the tension between human agency and the brute forces of nature. I begin by focusing on the various religious and cultural rituals that people have invoked in attempts to modify the weather. As I am not a cultural historian, the objective of recounting these cultural practices is not to cover them in painstaking detail, but rather to extract from them observations about the underlying assumptions that guide such thinking. From there I look at more contemporary efforts to develop the science of pluviculture. And finally, I close with a string of intuition pumps that explore many of our intuitions about what counts, and what does not count, as controlling the weather.

Rainmakers: Influencing the Gods

Greek mythology offers us the story of Helios, god of the sun, whose supernatural raison d'être was to pilot his chariot across the skies, from dawn until dusk, casting light across the land. Like many gods of the time, Helios was given to salacious philandering on the Earth, and through his trysts with the Oceanid Clymene, bore the illegitimate child Phaeton. Also like many gods of time, Helios was something of an absentee father, and so Phaeton grew up forever in his shadow, among mortals, knowing that he was a demigod but never being able to prove it. Sometime around his late teens, Phaeton got it in his pubescent head that he could impress his friends by showing them that he was, in fact, child of a god. He approached his father, trepidatiously at first, like many teenagers, to ask to borrow the car keys; Helios, familiar with the difficulties of driving his chariot, refused. Phaeton eventually persuaded his father to allow him to drive the chariot just once across the sky, which he does to catastrophic effect. A mere demigod, he cannot control the horses that he is tasked to control, and ends up driving the chariot across the sky, scorching the earth, and causing fires and floods that affect life all over the planet. Among the many effects of his actions, his crashing of the chariot is said to have created a gash across the sky that tracks the Milky Way, transformed the continent of Africa into a desert, redirected the course of the Nile, and made the skin of people in Ethiopia black.

There's a lot going on in this myth, and really too much to talk about in this section. Notice, however, the prevalence of environmental determinism again. Phaeton's crashing of the chariot is responsible for causing different races, creating differences across the land. But also notice that the gods themselves, powerful as they are, have difficulty controlling the weather, though they may be just powerful enough to tame it. At least in this myth, there seems to be a tension here between whether the gods *are* the weather or whether the weather stands apart from the gods, just as unbridled horses stand apart from a chariot. More to the point, there's the additional lesson that controlling the weather is not a feat for a man. It's not the kind of thing that a mere mortal should toy with, lest the consequences be catastrophic.

Throughout the book, I have mentioned but not gone into much detail on the various myths and religions that have been constructed around the weather. Among the observations I've made, I've noted that many cultures and people have interpreted the forces of nature as the intentional actions of a god or the gods.

From this it follows that if the gods are in the business of changing the weather or directing it in a way that fulfills their objectives, then the most direct pathway to controlling the weather will be to persuade the gods that they should be more beneficent. Indeed, persuasion directly to the good graces of the gods – through prayer, faith, offerings, sacrifice, etc. – has been one of the primary mechanisms through which people at various times and places throughout the history of the planet have tried to change the weather.

Prayer and rainmaking rituals have been a feature of social life since at least as long ago as the Neolithic period. There is considerable evidence to suggest that African and Asian pastoralists have integrated rainmaking ceremonies into their cultural practices. So too in the Americas. In the Mayan civilization there were various important shamans who, rather than departments of agriculture and commerce, would manage communication with the water gods or sun gods. At the same time, in North America, the Zuni, the Osage, and the Quapaw all had various rain dances that they performed to try to bring on the rain.[2] Tribes of Indians[3] across the Americas were involved in rainmaking rituals.[4] The Shasta and Wintu tribes would burn the splinters of trees that had been struck by lightning. The Patwin would burn the nests of diving birds. The Pomo Indians would throw ashes in the air to try to create clouds, whereas the Yuma would run in circles to try to create huge dust storms that would stir up the clouds. Often tobacco smoking was an important part of these rituals, ostensibly with the objective of making clouds.

Other tribes aimed to replicate thunder. The Ilahita Arapesh peoples of New Guinea would use flat, carved sticks called "bullroarers" to try to replicate the sound of thunder, as did the Hopi, Pomo, Patwin, and Washoe. The Yuki called their bullroarer "thunder's voice" and the Washoe called it a "thunder stick." Native Americans would also try to create the sounds of storms in an effort to conjure them. Other tribes would use thunder rocks and associated them with wind and rain.

But these are so far relatively innocent and innocuous practices. In some of the earliest civilizations, ritual sacrifice aimed to appease or please the gods, ostensibly with the objective of promoting agricultural fertility. Most historical records on sacrifice involve animal sacrifice, which was widely practiced in Ancient Greece and Rome.[5] Take this example from the *Odyssey* (430–446):

> So he spoke, and they all set busily to work. The heifer came from the plain and from the swift, shapely ship came the comrades of great-hearted Telemachus; the smith came, bearing in his hands his tools of bronze, the implements of his craft, anvil and hammer and well-made tongs, [435] wherewith he wrought the gold; and Athena came to accept the sacrifice. Then the old man, Nestor, the driver of chariots, gave gold, and the smith prepared it, and overlaid therewith the horns of the heifer, that the goddess might rejoice when she beheld the offering. And Stratius and goodly Echephron led the heifer by the horns, [440] and Aretus came from the chamber, bringing them water for the hands in a basin embossed with flowers, and in the other hand he held barley grains in a basket; and Thrasymedes, steadfast in fight, stood by, holding in his hands a sharp axe, to fell the heifer; and Perseus held the bowl for the blood. Then the old man, Nestor, driver of chariots, [445] began the opening rite of hand-washing and sprinkling with barley grains, and earnestly he prayed to Athena, cutting off as first offering the hair from the head, and casting it into the fire.[6]

The Romans partook of the *aquaelicium* (which in Latin means "calling the waters") and made offerings to Jupiter. But sacrifices were far from limited to nonhuman animals. Human sacrifice was common in Aztec culture, where children would be sacrificed to appease the rain god Tláloc.[7] The Mayans, famously, would sacrifice both prisoners of war and high-status members of their own civilization, sometimes by decapitation, heart removal, arrow sacrifice, and bloodletting from the ears, cheeks, lips, tongues, and penises of victims.[8] Nor were such practices limited to ancient history. Archaeologists have found evidence that, as recently as the 1400s, the Chimú civilization in Peru sacrificed hundreds of children and animals, apparently in response to El Niño conditions.[9]

The idea of these practices hinges on the supposition that there is some divine or external force that controls the weather. This is a natural way of making sense of forces that not only have a deep and important effect on us but also in many respects seem otherwise pointless and at times cruel. How else to explain the loss of our crops after we have put in so many hours working the fields? The gods must be angry. How else to explain the impact on our families as we are flooded out of our homes? The gods must be spiteful. If only there were some way to influence those who make decisions about when to send rain or sun.

It is a profound irony that it was likely massive weather disasters that eventually brought down almost all of these prior civilizations.[10] So much for weather control.

The Teleological Model

Since at least as long as Plato and Aristotle, and likely long before that, people have been thinking about the world around them as imbued with purposes. The basic idea behind this way of thinking is that all things in nature have some kind of *purpose*, that the stuff in the world is here for a reason. So, for instance, they might think that the purpose of rain is to water crops, the purpose of wind is to distribute pollen, or the purpose of clouds to shade the land. This is an idea in philosophy sometimes called "teleology" (from the Greek word *telos* (τέλος), which means "goal" or "purpose"). In this usage, *teleology* just means "the logic of purposes."

Teleological thinking is probably familiar to most of us in a more contemporary religious context, because it sometimes invokes the idea of intelligent design, suggesting that those natural purposes were put there intentionally by a supernatural entity. For instance, the Book of Genesis opens with precisely this framing:

> Gen 1.
> "In the beginning God created the heaven and the earth.
> "And the earth was without form, and void; and darkness was upon the face of the deep. And the Spirit of God moved upon the face of the waters.
> "And God said, Let there be light: and there was light.
> "And God saw the light, that it was good: and God divided the light from the darkness." ...
> "And God said, Behold, I have given you every herb bearing seed, which is upon the face of all the earth, and every tree, in the which is the fruit of a tree yielding seed; to you it shall be for meat."

> "And to every beast of the earth, and to every fowl of the air, and to every thing that creepeth upon the earth, wherein there is life, I have given every green herb for meat: and it was so."

Note here that God imbues the world and all its aspects with purposes. He creates these things, gives them a name, gives them a purpose, and then calls it good. The fruits of trees are there to feed humans and beasts. That is their purpose. The rain is there to feed the trees and the grass. That is what he *intended* rain to do.

It is important to see, though, that not all teleological views are equal. Purposes needn't necessarily be tied to a singular god (monotheism). They could, for instance, have been given by multiple gods (polytheism). Nor are these purposes necessarily even tied to supernatural intent, as it is equally plausible to propose that the purposes are "just there" in nature, without any kind of divine design.

Throughout the *Odyssey*, for instance, the teleology is more polytheistic, meaning that there were many gods meddling with the natural world. As we've already discussed, the Greek gods control the weather (or just *are* the weather). Zeus sends rain and lightning to assault Odysseus wherever he goes. Poseidon and Athena push ships along with storms and winds and waves. Even the nymph Calypso creates waves. The movement of the winds determines Odysseus's journey, pushing him in the directions that ultimately make for the entire backbone of the text. Prometheus was punished by Zeus for giving fire to the humans. This punishment was devised by Kratos, assistant to Zeus, and serves as the root word of democracy [δημοκρατία: *dêmos* (δῆμος) + *krátos* (κράτος)]. Kratos was "power."

For Greek philosophers like Aristotle, nature or *physis* (φύσις) had its own, more natural cause. In fact, Aristotle's view was mostly naturalistic, though it retained this teleological emphasis. It is the naturalistic and functionalistic view that most philosophers learn as independent of more theological interpretations. Plato, who preceded Aristotle, had a much more anthropocentric and creationistic view of teleology. Plato's thought was that the demiurge[11] – which is to say, the fashioner or shaper of the world (so not quite the monotheistic god that many Judeo-Christian interpreters might assume, but instead a secondary figure adjacent to the realm of forms) – was behind the activities of the world.

Teleological thought of the monotheistic variety was carried forward throughout the Middle Ages with work by important figures like St. Augustine (354–430), Boethius (480–524), Michael Scot (1175–1232), Albertus Magnus (c. 1200–1280), and St. Thomas Aquinas (1225–1274), all of whom added a distinctively theological tinge to the otherwise naturalistic teleology of Aristotle.[12] What I mean when I say this is that intent and purpose were sewn back into natural events, such that their *telos* could only be understood by appeal to the supernatural creator. The dominance of this monotheistic teleological model persisted throughout the late Middle Ages, just as most of western Europe was being hit with unexpectedly difficult weather: floods, pestilence, and even the Black Death.

And these views, or at least views like them, persisted for a long time. Aquinas himself would write of storms as diabolical in nature:

> Rains and winds, and whatsoever occurs by local impulse alone, can be caused by demon. It is a dogma of faith that the demons can produce wind, storms, and a rain of fire from heaven.[13]

Where earlier teleological stances tended toward attributions of intention to the gods, and thus sought to appease the gods, or at least plead to them, to make the weather good, the monotheistic orientation of Judeo-Christian thinking during the medieval period largely saw God as beneficent. This meant that good weather could be attributed to God, where bad weather would instead be attributed to demonic or diabolical forces. (Here again we see our division between the productive and destructive impacts of weather rearing its head, albeit instantiated in an effort to try to control the weather by rooting out demonic forces.)

During the Dark Ages, a secret school of magicians called Tempestarii were rumored to be capable of controlling the weather. They claimed that they could create rain or prevent storms and gained reputations throughout villages. In truth they were a kind of proto-huckster, urging commonfolk to pay tithes to the church as a primitive form of insurance. A notoriously anti-Semitic archbishop, Agobard of Lyon (c. 779–840), wrote a well-known screed railing against weather magic in which he accuses the Tempestarii of orchestrating a campaign to steal grain and bring it back to the magical cloud-land of Magonia. (Consider again how that relates to our earlier discussion of risk and risk pooling.)

The so-called Medieval Warm Period was cut short by the onset of the Little Ice Age in the early 1300s, which is widely attributed with having shaken up European history and alleged by some scholars to have coincided with the rise of witch trials.[14] Many of the medieval witch hunts were attempts to flush out witches who were thought to be capable of controlling storms and lightning, thereby preventing diabolical forces from causing bad weather. The *Malleus Malleficarum* (The Hammer of Witches, published in 1486), was on one hand a straightforward compendium of demonology, suggesting torture and death as a solution to witchcraft, but on the other hand also an effort to suggest that humans, practicing witchcraft, could use their powers to control the weather.[15] (Shakespeare wrote his *Tempest* 150 years later, in which the sorcerer Prospero creates a storm that shipwrecks his brother, Antonio, on an island, ostensibly for betraying him and staging a coup many years prior.)

In the 1484 papal bull, included as a preface to the *Malleus*, Pope Innocent VIII suggests that witches are capable of destroying crops by raising up hailstorms and tempests, causing lightning to blast men and beasts.[16] Inasmuch as these were mostly teleological views with a theological edge, they were intentionalistic in nature. So the strategy for trying to control the weather or manage impending weather events would be to change the intentions of the supernatural entities that were in charge of determining what the ultimate goal or *telos* of a given weather event would be.

Things eventually settled down as the Renaissance and Enlightenment came along, though intentionalistic, teleological, or theological interpretations of weather events never really went away. (I mean here that the Aristotelian or Thomistic worldview that had otherwise informed so much thinking about meteorology fell out of favor. The witch trials continued for a shockingly long time, resulting in over a million executions between the thirteenth and nineteenth centuries.)

No doubt countless natural disasters have been and continue to be attributed to the wrath of the gods, in Western society as well as many other nonindustrial societies.[17] The Great Flood of 1703 in Britain was said to be punishment for the depravities of London.[18] An 1886 earthquake in South Carolina led many folks to fear "judgement day."[19] Books and pamphlets with titles like *Global Warming or God's Warning* or *Acts of God* continue to populate bookshelves.[20] So too with ideas of "intercession" – which is essentially just the act of praying to a deity to get them to do something. These remain pretty common ways of trying to control the weather, so much so that they reach high levels of governance. In 2011, for instance, Texas Governor Rick Perry established the Days of Prayer for Rain in the state of Texas.

Mechanistic and Other Models

With the insights of the Renaissance and the Enlightenment, scientists and philosophers began a new tack on an old objective. Instead of appealing to the gods to make rain – operating on the assumption that the cleanest pathway through which to control the weather was to shift the intentions of the gods – practitioners turned to the skies themselves. The key conceptual shift occurred in understanding weather phenomena as primarily *accidental* rather than *intentional*. This realignment to emphasize accident over intent, of course, was manifesting throughout the scientific establishment, as we have already seen.

Figures such as René Descartes, Francis Bacon, Baruch Spinoza, and John Locke rose to prominence and would overturn some of the most important teleological beliefs. This intellectual movement introduced a new tension in meteorology, possibly brought about precisely because the weather had changed so dramatically. Features of the weather that otherwise seemed constant and static could suddenly be explained as incidental or accidental.[21] Francis Bacon's contributions to the development of the modern sciences were many, but even his more politically oriented work, *The New Atlantis*, played an important role in this. Bacon imagines a future utopia and mentions near the end of the book several technologies aimed to benefit mankind, including the "raising of tempests."[22]

Darwin's observations about natural selection operate according to a similar sort of logic. Since people stopped seeing the world as shaped by the intention of the gods but rather through physical forces and pressures within various kingdoms, a whole new strategy for controlling the weather was called for. Just as meteorological science came on the scene as a way of predicting the weather, so too came a good deal of information about the formation of weather systems and the physical inputs that make them function. And just as with the various conceptualizations that inform our explanations, each of these conceptualizations has slowly migrated through our thinking about how to control the weather.

Suddenly stripped of their purpose-oriented nature, the teleological model eventually ceded ground to a raft of different models altogether, all of which aim to explain weather largely on metaphorical grounds. The easiest to understand may be attempts to explain weather as analogous with machines. This "mechanistic model" proposes that weather systems are essentially physical and machine-like. Naturalistic and mechanistic ideas about the natural realm grew prominent during the industrial age, reflecting the proliferation of machines that were at the time racing across continents: steam engines, steam ships, trains, and eventually planes and automobiles. The central idea suddenly in play was that the atmosphere and the sundry weather systems associated with the atmosphere function like machines.

On the mechanistic view it seems obvious that the way to control weather systems is to figure out the right inputs and outputs so as to make them do our bidding, much like we might add gasoline to our car. Put in some fuel, out comes a product. "Mechanism" was not the only metaphor to inform scientific thinking about weather systems. There were other models to come out of the sciences as well. The biological/organistic model proposed that weather systems function less like a machine and more like a biological system or an organism.[23] On this view, one might feed or starve a storm, much like one feeds or starves a plant, and though the process by which weather functions might be hidden from view, weather systems might be understood to live or die on their food sources. Alternatively, a systems model might propose that instead of analogizing weather to a machine or to an organism, one should instead approach weather as a kind of system, in which all elements of the system function together. Other such models included statistical models, based on the suppositions that numerical probabilities could best describe weather, or fluid mechanics models, which aimed instead to understand weather systems as the confluence of fluids. Each of these various models would speak to a slightly different approach to modifying and controlling the weather.

Pluviculturists: Pulling the Knobs and Levers

The tradition of pluviculture[24] – basically, the "science" of rainmaking – emerged at the end of the nineteenth and the beginning of the twentieth century as a sort of middle ground between the teleological and more "scientific" approaches to weather modification.

Some pluvicultural efforts were rooted in sound scientific thinking, but much of it was buncombe. For instance, a 1920 issue of the *Bulletin of the American Meteorological Society* raises concerns about Charles Hatfield, who claimed that he could control the rain.[25] He sold his project to unwitting rubes on a gambler's proposition: he would receive no compensation for the first inch of rain, but should more than an inch fall, he would receive

$3,000 for the second inch and $3,000 for the third inch. It takes almost no sophisticated mathematics to see what a scam this is, predicated on the same mysteries of the weather that we were discussing earlier. Nevertheless, Hatfield was able to convince the city of San Diego, California to hire him to fill the Morena Dam reservoir with rainwater. This story would later provide inspiration for the 1956 Burt Lancaster film *The Rainmaker* and ultimately the song "Hatfield" by Widespread Panic.

Many projects preyed on the statistics of risk that we have been talking about throughout. Here's a sort of protoscientific effort from 1925 to lay out in detail a technique not only for making rain but for financing it as well:

> The pluviculturist has next to build a modest shack or to set up a tent for his chemical operations. Next he prepares certain chemicals in accordance with a secret formula. These may cost $50 more or less, according to the likelihood of further demands for extension of his operations. What the formula is, naturally no one has explained. Let me suggest a formula of my own. Take first ten pounds of pulverized chlorate of potash, and an equal amount of granular cane sugar. Mix these carefully in a wooden tub and when ready pour over them a liter (or pint) of sulphuric acid (e. p.). This simple and inexpensive preparation will produce surprising results. These may be brilliantly enhanced by using a pound of magnesium ribbon, to one end of which a lighted match has been applied, the whole sent into the air by attachment to a sky-rocket. This is most effective towards night or after clouds begin to form. Then certain salts of strontium yielding red light, barium yielding green, and

other salts yielding lights of different colors, should be set on fire. That this formula of mine has been used by any professional rain-maker, I do not know. I am sure that any pharmacist might furnish something equally good. Some also use an old-fashioned fanning, mill to condense the air, but that is less impressive.

Now that the chemistry has been provided for, the most important point follows, the economies of the process. There is an international institution known as "Lloyds" which insures anybody against anything, after a study statistical or meteorological of the chances. It charges a modest premium which naturally varies with the probabilities. If you want a clear day for a picnic, or a football game, Lloyds will for a consideration insure you against rain. Lloyds do not control the weather, but while losing the premium charged you will receive enough to finance your pleasure or your sport next time. You can insure a base-ball player against striking out, or an airship from falling into the sea, in accordance with scientifically accepted probabilities. Every well regulated stadium or other center of culture is a client of Lloyds.[26]

As during antiquity, people had all sorts of wacky ideas about what caused weather disasters, all of which informed the snake oil scams that were designed to mollify worried farmers. To explain the Dust Bowl, for instance, some in the 1930s thought that radio stations were causing drought; others thought, perhaps not erroneously, that the poisonous fumes from industrial production, cars, and planes might be the culprit; and still others thought that there was too much "vibration" associated with modern living.[27] And in many ways, strategies and hoaxes aimed at manipulating the weather morphed from their original supernatural upshot to take advantage of these unsupported theories. Where in medieval France, wine growers would use church bells to try to ward off the hail, the new science of pluviculture developed hail cannons that aimed not so much to frustrate tempests but to break up hail before it destroyed crops. These devices would shoot acoustic blasts into the sky every few seconds during a storm, hoping to agitate the air and break up hail before it formed.[28]

Box 10 Stinging in the Rain

A classic con used by alleged rainmakers exploited the probabilities of rainfall. Much like the psychics who trick unwitting victims into betting money on baseball games by sending out thousands of postcards predicting the winner of the world series, only slowly paring back their audience as their "predictions don't come true," fraudsters could make money by predicting the future to a large enough audience.

Scientists have tried a wide range of mechanisms to change the weather. Silver iodide is perhaps the best-known cloud-seeding chemical concoction, but others use dry ice, potassium iodide, lead iodide, liquid propane, and even table salt. The key idea was to inject a hygroscopic substance into the sky that will attract water molecules and extract them from water dense clouds. Other efforts at weather modification include the science of fog lifting, lightning suppression, hurricane snuffing, and snowpack enhancement. Some have suggested that depositing environmentally friendly oils or surfactants onto the ocean's surface will prevent evaporation, and therefore the formation of damaging hurricanes.[29] Likewise, enthusiastic entrepreneurs have been selling the weather equivalent of snake oil, calling to mind the cartoonish science fiction of the *Jetsons*. Dyn-O-Mat Company, for instance, developed the polymer Dyn-O-Gel – purported to absorb over 1,500 times its weight in water – which would allegedly prevent strong hurricanes.[30]

One of the more scientifically grounded pioneers in this regard in the 1940s and 1950s was fiction author Kurt Vonnegut's older brother Bernard. Some of this research is parodied in *Cat's Cradle*, the younger Vonnegut's novel about a substance called ice-nine that was developed by the military – a variant of water that stays solid at room temperature but that will, upon contact with liquid water, turn that liquid water into more ice-nine.

From 1962 to 1983, the US government funded Project Stormfury, which sought to weaken tropical cyclones by seeding the center of the storm with silver iodide.[31] The idea was to disrupt the structure of the storm and prevent it from causing extensive damage. Experimentation occurred throughout this period, including field studies and the development of numerical hurricane models. In 1969 the scientists were able to run tests on Hurricane Debbie, the results of which were promising and touted by scientists as a success, but ultimately they were inconclusive.[32] Despite the ambitious goals and extensive research, by 1977 the project faced significant political problems and it was scrapped as being unfeasible.

In Chapter 4, I argued that our experience with the weather is shaped dramatically by the way in which we conceptualize it. I spent some time discussing the idea that the statistical tools that we use to frame our thinking about natural events like weather go a long way in mollifying our approach to it. But as we can see here, our conceptualization of the weather is also paired with our implementation of technologies that build off of our advanced understanding of the world. Recognizing this, we can see that our advanced understanding of weather and our increasing capacity to manipulate it comes with a hefty reductionistic price tag – that is, that our attitudes about and dispositional orientation toward weather themselves reorient.

Reintroducing Intent

Think back now to Chapter 1, where I mentioned the "pathetic fallacy." I explained it as a literary device used to personify or attribute personal intentions to inanimate or nonhuman objects, born of sentimentalism and romanticism about the natural world. It is often used pejoratively by critics to suggest that commentators are being unrealistic or overly romantic in their assessment of real-world matters. In many ways it invokes some of the teleological thinking that I talked about at the beginning of this chapter.

But the philosopher Daniel Dennett has a nice way of thinking about this. Dennett coined the term "intentional stance"[33] to describe an attitude that we can take with regard to the natural world in order to understand it better. He argued there that we can try to explain or even predict the behavior of any object or natural phenomenon by assuming at least one of three different stances: a physical stance, a design stance (what I have here called the teleological model), or an intentional stance. He suggested in that book that it is a mistake to discount the attribution of intention to inanimate objects in order to make better sense of them. If someone says, "Lightning wants to get to the ground," what they mean by this is not literally that lightning has wants or intentions but only that lightning is attracted to the ground. The intentional stance thus does practical work for us.

If there's one thing that those who seek to control the weather aim to do it's to inject their own intentions into otherwise chaotic systems, essentially reintroducing purpose into those systems. The injection of our own intentions into the creation of weather then introduces additional concerns about the intentions themselves. That is, if it's the case that we somehow do develop the technological capacity to seize control of nature and put it to uses that we intend, then we will have to do considerably more work to explain or justify what we have done than we have in the past.

It is one thing for a rainstorm or a deep freeze to lay waste to our fields as a matter of historical or natural contingency; it is something very different for that same rainstorm or deep freeze to lay waste to our fields as a mistaken or hostile action by humans in control. It is enticing to believe that technologists and policymakers will use such technologies almost entirely for beneficial purposes, but we have no reason to believe that it wouldn't also be militarized and put to hostile purposes.

When the Japanese discovered the jet stream in the 1920s, for instance, they didn't naturally apply it to predictions of weather but immediately to militaristic uses. In World War II they attempted to send thousands of "Fu-go" hydrogen balloons toward the United States to light fires in the forests of the Pacific northwest.

At roughly the same time that Project Stormfury was underway in the United States, at the height of the Vietnam War, a highly classified cloud-seeding project code-named Operation Popeye was being conducted by the US military over the skies of Thailand, Cambodia, Laos, and Vietnam. The primary objective of this operation was to extend the monsoon season to disrupt North Vietnamese military supplies by softening road surfaces and causing landslides. From 1967 to 1972, military planes would disperse silver iodide and lead iodide into the clouds above the Ho Chi Minh Trail. A second, related weather modification effort, Operation Commando Lava, used planes to drop a specially engineered powder that would make mud and dirt more liquid, thereby hampering the progress of ground-based troops.[34] Both operations flew under the slogan, "Make mud, not war." Though the operations continued for roughly five years, once they were revealed to the public, controversy over the ethical implications of weather modification for military purposes began to swirl.

In the ethics literature this is sometimes called the problem of "dual use." The idea is intuitive enough: some technologies have two uses, one beneficial, the other harmful. Technologies that fall in this category include, for instance, nuclear, biotech, nanotech, artificial intelligence, surveillance, and drone. In the nuclear realm, obviously it's the case that we can generate a fair bit of electricity using nuclear know-how, but it's also the case that we can cause significant destruction in the form of nuclear weapons. So too with biotechnologies, where we might be able to innovate new medical cures but also cause considerable ecological damage by manipulating the genetic makeup of organisms and species. The same goes for nanotechnologies, where we might be able to develop new techniques for cleaning up chemical spills, while also generating hazardous small-scale environmental pollutants. As good global citizens we want to be careful to allow for beneficial uses but also restrict any harmful uses.

Acknowledging the dual-use capacity of these technologies is one thing, but regulating one use over another is quite a bit more challenging, as illustrated most notoriously by US gun policy. Guns are said to have both beneficial uses (they can be used for hunting or protection) and malicious uses (they can be used to kill other people). Try regulating gun use in the United States. The issue is so thorny that it is often said to be the third rail of politics.

The dual-use problem applies to weather control as well, as although it could be used to elevate the productive power of weather, it would likely also be used to harness the destructive power of weather. There are even international treaties that specifically outlaw the use of weather control for military purposes. The UN Convention on the Prohibition of Military or Any Other Hostile Use of Environmental Modification Techniques (ENMOD) was signed and ratified in 1977.[35] It entered into force in October 1978. Though not limited in scope to weather, it regulates the use of any environmental modification technique for hostile purposes.

We have already briefly discussed the extent to which weather events can create difficulties for military endeavors, so much so that they begin to shape history. If these weather events are suddenly under the control of human actors, they not only invite questions about justifications for their deployment but also threaten to upend tense and fragile relationships between states. This is why ENMOD is such an important document.

The problem is that all actors are not created equal. States are the natural signatories to treaties such as ENMOD, and they are naturally implicated in the deployment of weather control. However, if any of these technologies become more accessible, as has happened with most of our technologies, there's no reason to think that hostile uses are limited to the interests of states and their militaries. For example, an extremely powerful private industry could determine that it is in their interest to mess with economic markets. Industry actors are notoriously self-interested, almost by design, and if weather control is within arm's reach of states then it could also be deployed by a company hoping to undermine the competition or gain a leg up on the market. The same might be said about narcissistic or megalomaniacal private individuals.

You may further recall that the dual uses of the nuclear age were actually what inspired Ulrich Beck and Anthony Giddens to develop their idea of the risk society, and to note that many of our technologies engender new, manufactured risks. And that's part of the reason I introduced this problem here. As we develop greater and more powerful technologies, conceivably to control our weather, not only is there no guarantee that the purposes to which these technologies are put will be beneficial, but there is also considerable risk that private individuals and corporations may co-opt their development to further their own private ends.

Throughout the book, and at the beginning of this chapter, I have been discussing not just the ways in which we aim to control the weather but how we try to make sense of the weather and bend it to do our bidding. The issue that has been looming in the background, of course, is that of hazards and luck, of risk and risk management, of the sundry ways in which our intellectual tools for making sense of the world have given rise to abstract technologies that enable us to manipulate our relationship to weather. That is, there is a way in which we try to control the weather, not by manipulating the natural resources of nature – the light, water, snow, clouds, or wind – but by utilizing the tools of statistics and probability to control how weather affects us. We are able to predict, manage, control, and live with weather by utilizing all of the aforementioned resources.

The Temptation of Control and the Folly of Power

We are currently in the midst of the fourth agricultural revolution – primarily technological in nature.[36] We are now genetically modifying plants so that they are more weather hardy. We have developed crops that are more pest resistant, obviating the need for general applications of chemical fertilizers and pesticides. We have built and deployed expansive hydroponic operations that replicate ideal weather conditions, stacking them in vertical farming arrays so that they are easy to access using robotics, so that artificial sunlight can be fed to them from all sides for the appropriate number of hours. The heavy machinery of the third agricultural revolution is continuously being replaced by more intelligent robotic farmers that can monitor nutrient uptake from crops and tailor artificial weather to the needs of the plant. Robotics technology is already being used to plant, weed, fertilize, and harvest crops, but its role in agriculture will only grow more essential in the coming years. Artificial intelligence monitoring technologies can steer autonomous robots, identify disease, and anticipate and respond more intelligently to the weather. Cell-cultured meats are also coming on the scene, posing threats to conventional animal agriculture as well as vegetable-based substitute meats.

The fourth agricultural revolution promises near total control of the variables that have otherwise made agriculture a luck-bound endeavor. Though it is true that people have for centuries sought to control the weather, and to some extent have found different techniques for redirecting precipitation and the impacts of weather to embrace, resist, or harness weather, as we enter this next agricultural revolution – rife as it is with mind-bending technologies that only fifty years ago would have been the content of science fiction – it is worth asking what it really means to "control the weather."

There is, of course, the loose sense of "controlling" the weather that we covered in Chapter 3 related to living with the impacts of weather – accepting it, resisting it, or harnessing it – but most of the time, when we're talking about controlling the weather, this isn't what we mean. What we mean is that we will be able to control when rain comes and goes, how much precipitation is delivered and to where, how intense the sunlight or heat is, whether winds are destructive or gentle – and to do so in such a way that it works in our service rather than against us. We want weather to cooperate with our intentions. We want, in effect, to wrest intention from God. But even this explanation seems so nondescript and noncommittal as to leave us thinking that we *already* control the weather.

Now is actually a good time to pull another trick out of my philosophical cap and deploy a set of devices philosophers sometimes call "intuition pumps." Here I ask where the qualitative distinction between the two different senses of "controlling the weather" lies. Consider the following:

> **Sprinkler:** If Abercrombie sprinkles water over his crops using a sprinkler system that replicates the natural cycle of rain, is this a form of weather control?

It would seem on one hand that it clearly is not. Abercrombie is doing nothing more than channeling water from one location to another using a reservoir, a hose, and a sprinkler head. That can't be what weather control is. This looks much more like harnessing the weather, as we discussed earlier. On the other hand, how could this not be weather control? Is it not the case that the weather (in this case rain) is just water that has been transferred from one terrestrial surface to another terrestrial surface via the atmosphere?

Indeed, the ski resorts near me have gotten quite good at creating snow to extend the ski season and to increase the base layer. To do this, they typically just pump water out of a local reservoir that has accumulated water over the summer, spray it into the sky when the temperature is below $27.5°F$ ($-2.5°C$), and snow falls on the slopes. They call these efforts "snowmaking," which sounds a lot like *controlling* the weather, though in practice it looks a lot more like *harnessing* the weather. Surely controlling the rain must be akin to making snow. Is there a categorical difference we're missing here? Some important difference between injecting a catalytic substance into the sky, like silver iodide, to inspire snowflake formation, versus injecting *water* into the sky, to similarly inspire snowflake formation?

It seems at least plausible that if we are to say that we genuinely control the weather and don't just harness it, we would want to say that we are *extracting* the precipitation from the atmosphere, rather than just moving water around. But consider:

> **Blower:** If Bartholomew installs an enormous fan on one side of a mountain to push a cloud to the other side of the mountain, redirecting the cloud to sprinkle water again over crops in the valley on the other side of his mountain, is this controlling the weather?

On one hand, it seems to be more directly controlling the weather, as it will be raining in one location while it will not be raining in another location, and all Bartholomew is doing is moving the cloud from a less desirable place to another desirable place. On the other hand, it doesn't seem to be importantly different than the sprinkler case of harnessing the weather, as we discussed earlier. A cloud is essentially just a sky reservoir, so what should it matter whether the water is in a ground-based reservoir or in the sky? It seems to be instead another case of steering the rainfall, redirecting the drops so that they land in a more specific location. There's something to this, in that controlling the *location* of weather matters, and to have control over the location where rain falls would take us a long way in saying that we control the weather, but there are still quite a few additional complications.

For such a strategy to work, Bartholomew would have to wait for a cloud to materialize. Without a cloud, he'd have no rain to redirect. He'd also have to be sure that other systems could not overpower his fans, as a single errant wind might blow his clouds off target. And he'd further have to come up with a way of turning the rain off, should it be the case that a larger storm were to come in and swamp his clouds. He's still, in some respects, "at the mercy of" the weather.

So maybe there's something else to weather control. Maybe it has to do also with the *genesis* of the rain. To control weather, one has to be the source of the weather.

> **Evaporator:** If Claudius instead develops a contraption that evaporates water off of a nearby lake, creating a cloud that will, with the assistance of a giant fan, blow over his crops, and then introduces a trigger that will release the water from the cloud to rain on his crops, is this a form of weather control?

On first assessment, this would seem even more like what we mean when we say that someone is controlling the weather. Claudius controls the formation of rainclouds and governs the conduit through which the rain is conveyed to his crops. He has avoided the concern that he must wait around for the weather to cooperate. But is there really a qualitative difference between what Bartholomew does and what Claudius does?

Certainly there is an important sense in which Bartholomew is only pushing weather around whereas Claudius is generating precipitation by relying on the air to contain and redistribute water from elsewhere. But he's not really creating weather, any more than the ski slopes near me are creating snow. He's identified a source of water, has created a catchment for that water in clouds, and then redistributes that water somewhere else. That may seem like a minor point, but consider the problem of jurisdiction.

Assume, for instance, that the lake and all its contents belong to Farmer Jones, such that it is entirely possible that Claudius is in effect *stealing* the water from the nearby lake. (Or, alternatively, that the lake sits in a large jurisdictionally governed space, such as a state or a country.) By deliberately diverting water from the lake into the air and then onto his crops, Claudius looks to be redirecting water away from that jurisdiction. His control of the "weather" is really control of somebody else's water. What Farmer Jones once controlled, Claudius now controls, an action that is ripe for contestation and dispute. That of course is a form of control, but it's not really "control of the weather." Jones would have every reason to be quite angry about Claudius's illicit acquisition of his water. More than that, though, it looks like Claudius controls not the weather but Farmer Jones, and that can't be what controlling the weather is. Farmer Jones is not the weather.

So let's take the jurisdictional concerns out of the picture:

> **Ocean:** What if Dagney builds a contraption that evaporates water off of a large, unowned body of water –

the ocean, say – and then primes the clouds in such a way that they redirect their water from the ocean to his fields?

This seems to address the jurisdiction problem from before. It makes at least some difference that Dagney is taking water from the ocean to fill his clouds, whereas Claudius is taking from a lake (which has clear-cut boundaries and an implied jurisdiction). In other words, the ocean water, which was previously under the control of nature, now is under the control of Dagney, whereas Claudius has only seized control of Farmer Jones. It would seem, indeed, that jurisdiction, or prior control, matters when it comes to weather control. In order to control the weather, one must seize this control from *nature*, one must bend nature to do their bidding, and one must contend with the various other curveballs that nature throws in the way.

So let's try to prevent nature from wresting control back from those like Dagney who have taken his water. What if a sprinkler system is set up indoors, inside a warehouse, say, to replicate the world outside?

Suppose a high-tech weather replication system (call it the Weatherator 9000) inside a hermetically sealed environment (a Biosphere, say) informed and controlled by conditions external to the Biosphere. Suppose that there are sensors all around the exterior of the Biosphere – thermometers, barometers, hygrometers, anemometers (for measuring wind speed), disdrometers (for measuring drop size distribution), pyranometers (for measuring solar radiation) – and that these sensors collect information that is translated into numerical data that can be fed back to the Weatherator, which can then translate this data back into temperature, pressure, humidity, wind, etc.

> **Biosphere:** A farmer, Ernie, has programmed his Biosphere to replicate the weather immediately outside. Every time it rains two inches outside of the Biosphere, a sprinkler system will replicate the same amount of rain falling inside. The difference between the precipitation inside the building and outside the building is that Ernie, the farmer, maintains the possibility of turning the rain on and off at a time that he deems appropriate, though he never does so.

Given that Ernie has programmed the Weatherator to replicate the weather, and so in this respect the Weatherator only ever does just that, is it the case that Ernie controls the weather? It would seem that he is not controlling the weather, although he maintains the *possibility* of turning the Weatherator on or off, the machine is being guided and steered by the weather itself. This is a good start, because it suggests that there is at least some sense in which forces from outside the Biosphere – that is, the weather – won't interfere with forces inside the Biosphere, which Ernie controls. To control the weather, one must have the capacity to make the weather do what one intends.

> **Remote Control:** Now consider a different farmer, Finnegan, who has built a Biosphere in a remote part of the arctic but, given satellite technology, can download the same kinds of information that Ernie is using, albeit from anywhere on the planet. Using the Weatherator 9000 he can, from across the globe, recreate temperate conditions in his arctic Biosphere. Every time it rains two inches in the rich farmlands of Mexico it will also rain two inches inside Finnegan's Biosphere. Every time the temperature reaches 90°F, so too will temperatures reach that level inside

> Finnegan's Biosphere. The Weatherator 9000 is so sophisticated that it can do this virtually anywhere, even in the Arctic, so Finnegan can now grow crops nonindigenous to the region. Is Finnegan controlling the weather?

It would seem that, in a manner of speaking, he is. He has programmed his Weatherator to replicate weather that is conducive to growing crops in areas where they would not otherwise grow. His crops will know no difference. But there's still something that feels very much like a cheat here; it is not *exactly* like controlling the weather.

No doubt, this has something to do with the fact that Finnegan is doing this all indoors, essentially blocking out the actual weather and controlling only the conditions within the limited domain of the Biosphere. It seems to really matter that, in order to properly *control* the weather, one must actually *block* the weather. In this case, the weather is being blocked both by the walls of the Biosphere and by distance, as weather is being recorded from far away. The cheat occurs because Finnegan is only controlling the *inside weather*, where the *outside weather* remains outside of his control. So he both does control the weather (the inside weather) and doesn't control the weather (the outside weather). This would seem to suggest that he is not actually then controlling the weather.

> **Time Control:** Consider finally a different farmer, Gallagher, who has access to weather records from the year 1994, which was a particularly good harvest year. Suppose that there is a detailed historical record of rainfall, temperature, sun intensity, humidity, barometric pressure, wind, and so on, so that given this information,

> with the Weatherator he can replicate the weather perfectly. He can plug in his dataset to his high-tech sprinkler system and control an indoor farm to recreate the weather from the year 1994.

These are cases in which real-world weather has been recorded and reconstituted through technology, essentially played back as if by video. Here the weather is being blocked, not by distance, but by time. To control the weather, one must make it do what one intends in a specific place and time.

Minimally it seems like the conversion of weather and weather events into numbers, and then reconstitution in other locations, far from their original source, adds an element of distance between the recording of the event and its recreation. But there are a few things we know about recording real-world events and even real-world seasons. If we record our children on video and then watch this video when they are grown, we would scarcely want to suggest that we were controlling our children. In this instance we only control their image, which is actually just a bunch of pixels, coded as zeros and ones. We can manipulate that image, sure – adding elements, changing colors, moving things around – but we don't actually control them. So too with the Weatherator.

There is, as yet, no perfect season, even in the natural environments often considered ideal for planting. No system of weather patterns has ever met the Goldilocks ideal. There are days that are too hot, rainstorms that are too strong, periods of drought that linger too long. From the standpoint of agriculture, it would be much better to idealize conditions, perhaps by taking a more experimental approach and then using this data to recreate an ideal weather pattern for each crop. So it seems that the only real way to control the weather would be to control all aspects of the weather, and this seems only truly possible inside a hermetically sealed space like a Biosphere.

It would seem instead that weather is always and only an outside force that we cannot control. If we bring the weather indoors and create rain or change the temperature or add wind, that's all we're ever doing: manipulating variables that themselves are not weather but instead are aspects of what looks like weather. If we later bring this same notion of weather control outdoors and improve the regional temperature by pumping cold air into a valley, or by creating shade through an enormous scrim or a shield to block the sun, then we are still doing nothing much different than building a giant retention dam. We are not "controlling the weather" but only managing a variable of the weather.

And it is in this way that I think we will forever be chasing this white whale of trying to control the weather until the point at which we completely contain and prevent outside forces from manipulating whatever it is that we're trying to do. We can create entirely closed environments in which the elements of weather are managed with knobs and levers, in which the impacts of weather are managed through clever statistics, but we will never be in a position to suggest that this is a form of controlling the weather.

The dream of technocrats and supervillains will always and forever be out of reach.

******

To really understand the qualitative difference here it seems that we will need to lean back on our definition of weather from earlier. We established at the beginning of the book that weather is not precipitation but rather a force, and throughout the book we have slowly added to that conception. It seems again that this definition will serve us, since most of the cases that I've given in this chapter involve redirecting water and light to replicate conditions in the outside world. If weather is not a ledger of precipitation and conditions but instead an *outside* force, to control weather would entail controlling this force, taming that which is otherwise untamable.

Consider finally a considerably more fantastical case:

> **Total Control:** An evil supervillain, Boris Badenov, wrests control of nature and from his supervillain hideout begins to manipulate the clouds in the sky, the moisture in the air, and the wind around us.

We see this supervillain motif in popular culture all the time in comics, movies, novels, and cartoons. (Notably, we see almost no depictions of his superhero counterpart, Grisha Goodenov, a technology entrepreneur so powerful and beneficent that he uses the weather to help humanity.) There is a sense in which Boris may exemplify the ideal of weather manipulators everywhere: He controls the weather in every given respect. But there's another important way of looking at this scenario. This is the sense that we have been exploring for the past few chapters: the sense that construes weather as an outside force instead of a litany of events and phenomena. On this interpretation, Badenov's powers are a pure fiction, rooted in a mistaken conception of weather. The supervillain motif is more an outgrowth of the artifice of the scenario than it is truly a case of controlling the weather.

To explain: I have stipulated in this scenario that Boris Badenov, supervillain extraordinaire, has the powers of a god. Whatever he chooses to do, the weather – *defined* as those forces over which Badenov has control – follows. If instead I had stipulated that Badenov determines (1) how much rain will fall, (2) how hard the wind will blow, and (3) what the temperature will be, and continued numbering the many things that he has the power to do, then there will always be some remainder: some weather-related phenomena that he does not control. Perhaps the barometric pressure will not be in his control, or perhaps the amount of sunlight will be stronger than he had hoped. In this sense, Badenov would not be the supervillain that we have stipulated but would instead be more like Ernie, Finnegan, or Gallagher. We would be forced to admit that Badenov does not, in fact, control the weather but only controls some subset of weather-related phenomena. And this is not so unlike the situation we encounter with our current technologies.

More importantly, as passive observers in this scenario, *we* have no control of the weather, which is what enables us to talk using these fictions in the first place. Inasmuch as we have no control over the weather, there is an ineluctable aspect to weather control that keeps it always out of reach. That is, the forces that control the weather in this instance – Boris Badenov in this example, but the gods of mythology in prior examples – are always and forever alien to us. They are heteronomous.

So that's the last bit of our definitional puzzle. Weather is not the kind of force over which *we* human beings exert control; rather, it is a force that always already appears to be governed by powers outside of our control. The laws by which it abides are *heteronomous* in this way. (*Heteros* [ἕτερος] from the Greek meaning "other party" or "another" and *nomos* [νόμος] from the Greek meaning "law.") Heteronomous forces stand in sharp contrast to *autonomous* forces, since *autos* (αὐτός) relates instead to the self. When we say that we are autonomous, we mean that we are self-lawgiving, we act according to self-set purposes. When we say that some force is heteronomous, we mean that it acts according to other-set purposes. The forces of weather are governed by laws outside of us that remain, in many respects, invisible and mysterious.

We should, one last time, return to our working definition and see if there is more to say:

> Weather is a *heteronomous* force, chaotic but lawlike, both destructive and productive, civilization-shaping and anti-humanizing, that, coupled with human creativity, is an engine of innovation. In many cases it seems to have a mind of its own.

If what I'm saying here follows, that weather is a howling force always already around us, always surrounding us – Kairos, a fleeting, qualitative moment in time; Kratos, an alien power that refuses to be subdued – then we shall never be able to control the weather, for the weather is forever out of our control. It is *by definition* that which we cannot control. And that brings us back to the beginning of this chapter, where we discussed the means by which we might harness precipitation and the products of weather to achieve our own ends. All of our efforts at weather control only ever amount to attempts to harness the products of weather, to shift around the resources of the planet in such a way that they make our lives better.

Put differently: *It's not weather if we control it.* That's just what weather is – a force that we cannot control.

But, you may be inclined to object, what about more modest weather modification efforts? What if we could prevent large hurricanes or tornadoes from destroying places we live and love? If we could, for instance, make a hurricane into a gentle breeze, and we could do so without any concern that it would spin out of control. Would we not say then that we could control the weather? Maybe prevent massive destruction in the face of an event or a force that is much larger than us?

And I think, yes, in a manner of speaking, we could gain substantial control of the world around us, of nature, just as we have been doing with almost all of our technologies. But also, no, we would still not control the weather, because the weather will always and forever be that chaotic aspect of our lives that sits outside of our control.

The only reason we persist with the fiction that we can control the weather is because we misunderstand what weather is. Our cartoon villain – Boris Badenov – sits hidden away in his mountain lair, nefariously plotting to control the very systems upon which we depend. This cartoon villain can control the weather precisely because he is a character of make-believe, playing the role of a god. He might as well be Zeus or Kairos or Occasio; Perun, Indra, Leigong, or Ra.

Conclusion
Silver Linings

It is a standing joke in academia that some of the worst undergraduate papers begin with the phrase "Since the beginning of time ... " and then go downhill from there. I easily could've started this book off the same way. "Since the beginning of time, people have been talking about weather ... " But in this instance at least, people actually *have been* talking about weather since the beginning of time – seriously, likely at least since the moment that we could begin talking about anything – and they have been conjecturing and hypothesizing about how weather will affect them. All this to say, I can't purport to give a comprehensive overview of everything that's ever been said in a short book like this.

Nevertheless, I think we've done a fair bit of interesting work.

What I have tried to argue is that in our drive to quantify, operationalize, and scientize weather, to become more rigorous with our understanding of weather and weather events, to wrest control of nature through our advanced ways of knowing and doing, to develop ever more innovative technologies that replicate weather, we are also distancing ourselves from a fuller, more complete sense of weather. Much as we have advanced our understanding of and shifted our comportment toward weather, we have also, in some respects, taken a few steps backward. Our advancements in meteorology, improvements in weather prediction, innovations in risk management, and newfangled harnessing technologies have a way of masking the relational aspects of weather that make weather what it is. We are not necessarily learning all there is to know about weather but instead learning some specific things about weather and forgetting others, constraining how we understand our relationship to weather and in many ways how we understand our relationship to ourselves.

So let's quickly recall what we've covered.

In Chapter 1, I explored what weather is, investigating the metaphysics and ontology of weather's various manifestations. I began this exploration by raising familiar examples and then trying to bring these together to get at the concept behind weather. I examined many instances of weather – rain, snow, sleet, hail, thunder, lightning, clouds, sun, wind, storms, cold snaps, heat waves, clear skies, tornadoes, hurricanes, etc. – and discussed the ways in which these examples of weather ultimately fall short of offering a suitable definition. I also covered the ways in which metaphors of weather appear in literature, film, and popular culture, often as indications of tumult or unpredictability. The chapter bent toward a characterization of weather as a force that functions independently of our own willful activities.

In Chapter 2, I looked at the way in which weather has and does affect us, specifically establishing weather as both a *productive* and *destructive* force, but also ultimately

CONCLUSION

an *indifferent* force. I covered some of the moral categories that go into our assessment of impacts. I looked not only at our efforts to quantify the impacts but also at some of the less obvious qualitative shaping effects of weather. Specifically, I challenged the idea of "climate determinism" or "environmental determinism." The upshot of Chapter 2 was that weather isn't just an event-causing force but a force that affects us, and that inasmuch as it affects us, weather carries good and bad valences that we evaluate and build our lives around.

Having addressed the impacts and effects of weather on our lives, in Chapter 3 I discussed the various ways in which we've struggled to fight against or live with the weather. I framed this as an exploration of dispositional attitudes and suggested that the moral valence of weather is in part a consequence of the technologies and policies we have developed to mitigate risk. Roofs, gutters, aqueducts, pumps, shades, fabrics, paints, umbrellas, parasols, sunscreen, etc., have all done considerable work to dampen or amplify the impacts of weather on our lives. I also reflected on the three historically significant agricultural revolutions and tied them into the emergence of technologies and policies that we have used to intervene with weather. These technical innovations have themselves also shaped whole economies, transformed cities, and affected the physical landscape in which we live. I particularly stressed how contemporary theorists have sought to capture weather as one of many "ecosystem services," an actuarial abstraction that further reframes weather, not as an unending cascade of unpredictable hazards, but instead as a gift of free services from nature. This, I suggested, transforms our relationship to weather almost entirely into impact terms. My central purpose in this chapter was to make a *practical* point: Weather presents a kind of ongoing, forever-looming natural hazard, but as we've been able to soften the blow of weather through practical and technical means, we have changed how we live and how we view weather.

Where in Chapter 3 I focused on the practical implications of fighting against or learning to live with the weather, in Chapter 4 the discussion took a slightly more abstract turn. I discussed the move to modern meteorology and the science of weather. As meteorology moved from antiquity through modernity, as we've sliced and diced the various aspects of weather into measurable, quantifiable units, we have demystified and changed our thinking about weather altogether. Without question, this conceptual slicing and dicing has increased our understanding of weather phenomena and improved the predictive validity of our forecasts, but it has also in many ways removed us from the most hazardous front lines of weather. So my aim in this chapter was more *epistemic* than practical, to suggest that our relationship with weather has changed as we've learned to conceptualize weather differently. In the final section of the chapter I discussed the ways in which the demystification and quantification of weather has been adapted to characterize weather and its impacts as *risk*.

In Chapter 5, I addressed the tension between human agency and the brute forces of nature by exploring past and present attempts to control the weather. I began by

CONCLUSION

focusing on the various religious and cultural rituals that people have invoked in attempts to modify the weather. The objective of recounting these cultural practices was to extract from them observations about the underlying assumptions that guide such thinking. For instance, the idea that weather is an intentional force, steered by gods who may be listening; or, alternatively, the idea that nature is a mechanistic system that can, like a complicated thermostat, be adjusted to produce the right temperature. Bearing this in mind, Chapter 5 shifted to a series of intuition pumps, all aimed at suggesting that the forces of weather are always outside and alien – heteronomous – and that this heteronomy is encapsulated in the very idea of weather.

What I've argued essentially is that the scientific stance that many of us naturally take with regard to weather – in our meteorological endeavors, in our climate research, and in our daily exchanges about the forecast – has too narrowly defined its own subject, and in so doing it has failed to see the much bigger, much more interesting, philosophical picture.

Weather is an outside force that will forever push back on us. It has always done so. It will continue to do so. It provides the backdrop for our lives but also pushes us around in ways that shape who we are. Weather gives meaning to our day-to-day existence, provides events that delimit our actions, offers pushback when we push forward. It keeps our gardens and forests green and lush, it puts food on the table, it provides water for swimming, snow for skiing, ice for skating; and yet it also creates ever present hazards and dangers that threaten to wash away our homes, strand us in conditions we cannot endure, and starve our precious crops of sunlight and water.

Weather has historically served as the tumultuous intermediary between civilization and nature, first appearing in the garb of the gods, and then as the slave of the sciences, and ultimately as the jester of the Weather Channel, YouTube, and TikTok. Weather is the medium in which we decide what to do. It is a reminder that we are not in as much control of our lives as we often like to proclaim ourselves to be.

Over the course of this book, I have argued that weather is a powerful force that shapes who we are and how we live: a force that surrounds us and seems to have a mind of its own. As we have developed technologies to shield ourselves from weather, to predict its eventual twists and turns, and eventually to harness it for our own purposes, we have both changed our attitudes about weather and changed ourselves. Just as civilization is a response to weather, civilization has also separated us from the experience of weather, from the phenomenon of being tossed about by winds or having our family photos and memories dissolve in a flood. It has done so through two primary mechanisms: first, by finding ways to shield us directly; and second, by transforming the experience of weather into quantifiable metrics – temperature, humidity, barometric pressure, and the like.

We will always have weather, even if we get so sophisticated in our technologies of control that we can replicate weather in Biospheres. Why is this? Because whatever weather is, it is definitionally the kind of thing that is always already outside of our control. Weather is a heteronomous force that pushes back against us, that reminds us of our lack of control. It is fate. It is destiny. It is nature, red in tooth and claw.

If you agree, or at least have patiently slogged through this book, hopefully you won't find yourself in these uncomfortable office party circumstances anymore. You may instead imagine an office party that looks a bit more like this.

There you are, standing alone near the punch bowl. Over comes Clarissa, the intimidating albeit friendly manager who hired you. "Hi," she says, eyeing the punch bowl. "Hi," you reply. There is a moment of awkward silence. "Beautiful weather today," she says. "Yes," you nod. "They're saying sun again tomorrow," she says. And this is your entry point, your moment to say a few interesting things about what weather is, how it affects us, how we measure it, how we predict it, even how we have tried to control it.

Good luck. May Kairos – that is, weather – be your guide.

NOTES

Chapter 1

1 Orr G (2011). It's Raining Birds and Frogs. *The Independent*, January 6. www.independent.co.uk/climate-change/news/it-s-raining-birds-and-frogs-animal-phenomena-are-surprisingly-common-but-why-do-they-happen-2177017.html
2 Dennis J (2013) *It's Raining Frogs and Fishes: Four Seasons of Natural Phenomena and Oddities of the Sky*. Diversion Books.
3 The other fun quip is that Colorado has not four but twelve seasons: Winter, Fool's Spring, Second Winter, Spring of Deception, Third Winter, The Pollening, Actual Spring, Summer, Hell's Front Porch, False Fall, Second Summer, Actual Fall.
4 Izon G, Zerkle AL, Williford KH, Farquhar J, Poulton SW, and Claire MW (2017) Biological Regulation of Atmospheric Chemistry en Route to Planetary Oxygenation. *Proceedings of the National Academy of Sciences* 114(13), E2571–E2579.
5 Sagan C and Mullen G (1972) Earth and Mars: Evolution of Atmospheres and Surface Temperatures. *Science* 177(4043), 52–56.
6 Space.com Staff (2006) Jupiter Data Sheet. November 17, www.space.com/3134-jupiter-data-sheet.html
7 Tillman NT (2016) How Hot is Mercury? November 29, www.space.com/18645-mercury-temperature.html
8 Kraus D (2006) On Neptune it's Raining Diamonds. *American Scientist*, www.americanscientist.org/article/on-neptune-its-raining-diamonds
9 Putnam H (1973) Meaning and Reference. *The Journal of Philosophy* 70(19), 699–711.

10 Philosophers will recognize this as a very cartoonish version of Hilary Putnam's famous "Twin Earth" thought experiment. Mostly his is an argument about reference and meaning, but since we're talking about definitions of weather, it seems appropriate here.
11 Macauley D (2010) *Elemental Philosophy: Earth, Air, Fire, and Water as Environmental Ideas*. State University of New York Press.
12 If you trust the internet there's apparently no settled answer to this question. Quote Investigator (2012) The Climate is What You Expect; The Weather is What You Get. June 24, https://quoteinvestigator.com/2012/06/24/climate-vs-weather/
13 In the climate discourse it is common to hear that we have locked in two degrees of warming, and frequently this will be presented in the climate literature as "two degrees C," partly because the UN set a two-degree target in the Paris Agreement. To an American observer, however, accustomed as we are to using the Fahrenheit scale, "two degrees C" doesn't seem like much of an increment. A mere two degrees? You can go through a two-degree swing in half an hour barely even noticing the change. What really needs to be emphasized is that "two degrees C" is "3.6 degrees F," and that four degrees C, which is completely within the realm of possible climate futures, is a substantial increase above this: 7.2 degrees F. The point I'm making here is commonly taken as a point about communication, and about which metrics we should be relying on to get the most people to pay attention to climate change. But the observation here isn't so much about communication, or even about climate confusion, but rather that many of the metrics that we use – presumably to clarify and standardize different aspects of the weather in differing contexts – can and often do also have a masking effect. That is, they can serve to cloud or crowd out otherwise very important observations by shrouding them in numbers.

14 Juster N (2011) The Phantom Tollbooth, 50th anniversary edition. Knopf Books for Young Readers.
15 Translating of old English is difficult, and I am no expert, but this seems to mean something like, "And he brought in logs for a fire, for it was cold weather." Paulus Orosius, *Historiarum adversum paganos libri septem* (c. 416), emphasis added.
16 Given the above observations about time and weather, it's not a totally insane connection. Though the ideas are not linked etymologically, it makes at least some sense counterfactually. Try this little exercise: the next time somebody asks you what the weather is today, swap the words out in your head, substituting "whether" for "weather." I started doing this after exploring weather's connection to time. In my head, I began asking, "How's the whether?" If you understand weather in the way that I was understanding it, it brings to the fore the sense of momentariness, contingency, and flux that I think is actually captured by the idea of "weather."
17 Smith JE (1969) Time, Times, and the "Right Time": "Chronos" and "Kairos". *The Monist* 53(1), 1–13.
18 Aesop (2008) in Gibbs L (trans.) *Aesop's Fables*. Oxford University Press, 536.
19 I want to be up-front that I am no philologist and am a fair way out of my depth with regard to the origin of terms, but even if there is no direct etymological connection – from my research I uncovered some dispute among linguists, for instance, between *tempestas* and *tempus* – it is hard to escape the notion that speakers of the period would have felt these connections as the terms were used. Gzerga points out, for instance, that when the phrase "fair weather" was used in Old English, this was less likely to mean "calm" weather and more like to mean "right" weather, as in: the weather was right for sailing. Grzega J (2022) Climatic Conditions and Lexis: Some Diachronic Notes on Weather-Related Words in English and

Other European Languages. *Transactions of the Philological Society* 120(2), 320–331.
20 Strauss S and Orlove BS (2021) *Weather, Climate, Culture*. Routledge.
21 Knox B (1997) *The Odyssey*. Penguin.
22 Russell BW (2016) All the Gods of the World: Modern Maya Ritual in Yucatán, Mexico. *Ethnoarchaeology* 8(1), 4–29.
23 The phrase is often attributed to Edward Bulwer-Lytton's novel *Paul Clifford* as an example of how not to write. In the original, it reads: "It was a dark and stormy night; the rain fell in torrents – except at occasional intervals, when it was checked by a violent gust of wind which swept up the streets (for it is in London that our scene lies), rattling along the housetops, and fiercely agitating the scanty flame of the lamps that struggled against the darkness." Bulwer-Lytton E (1833) *Paul Clifford*. Baudry's European library.
24 Parrish SS (2012) Faulkner and the Outer Weather of 1927. *American Literary History* 24(1), 34–58.
25 McKim K (2013) *Cinema as Weather: Stylistic Screens and Atmospheric Change*. Routledge.
   Political Gabfest (2023) Trump Legal Traffic Jam. July 20, https://app.podscribe.ai/episode/86916664
26 Ibid.
27 Whitman C (2017) The Art of Making Your Own Weather. *Forbes*, October 6, www.forbes.com/sites/forbescoachescouncil/2017/10/06/the-art-of-making-your-own-weather/?sh=306c38c75c1b

Chapter 2

1 This is my preferred "head canon" reading of the Force and I will die on this hill. Rinzler JW (2013) So What the Heck are

Midi-Chlorians? www.starwars.com/news/so-what-the-heck-are-midi-chlorians
2. Juliano TW, Lareau N, Frediani ME, Shamsaei K, Eghdami M, Kosiba K, Wurman J, DeCastro A, Kosović B, and Ebrahimian H (2023) Toward a Better Understanding of Wildfire Behavior in the Wildland–Urban Interface: A Case Study of the 2021 Marshall Fire. *Geophysical Research Letters* 50(10), https://doi.org/10.1029/2022GL101557
3. Koppe C, Kovats S, Jendritzky G, and Menne B (2004) *Heat-Waves: Risks and Responses*. World Health Organization, Regional Office for Europe.
4. Weather-Related Deaths and Injuries. *National Safety Council Injury Facts.* https://injuryfacts.nsc.org/home-and-community/safety-topics/weather-related-deaths-and-injuries/data-details/
5. White House (2020o) The Rising Costs of Extreme Weather Events. September 1, www.whitehouse.gov/cea/written-materials/2022/09/01/the-rising-costs-of-extreme-weather-events/. When first accessed, this information was available on the primary webpage of the US White House. It has since been removed and can now, as of October 2025, be located at https://bidenwhitehouse.archives.gov/cea/written-materials/2022/09/01/the-rising-costs-of-extreme-weather-events/
6. *The Weather Channel* (2024) US Heat Deaths Soared to Record High Last Year. August 29, https.//weather.com/news/climate/news/2024-08-26-heat-deaths-by-year-in-us-hit-record
7. Chang LW, Kirgios EL, Mullainathan S, and Milkman KL (2024) Does Counting Change What Counts? Quantification Fixation Biases Decision-Making. *Proceedings of the National Academy of Sciences* 121(46), e2400215121, https://doi.org/doi:10.1073/pnas.2400215121
8. Smith AB and Katz RW (2013) US Billion-dollar Weather and Climate Disasters: Data Sources, Trends, Accuracy and Biases. *Springer Nature* 67, 387–410.

9 Social Determinants of Health (2024) *US Centers for Disease Control and Prevention.* January 17, www.cdc.gov/about/priorities/why-is-addressing-sdoh-important.html
10 Smith AB and Katz RW (2013) US Billion-Dollar Weather and Climate Disasters: Data Sources, Trends, Accuracy and Biases. *Natural Hazards* 67(2), 387–410.
11 Fun fact: Though the Dartmouth Flood Observatory is ostensibly connected to Dartmouth College and was indeed run out of the college from 1995 to 2010, it moved in 2010 to the University of Colorado Boulder and operates out of a lab just down the hall from my office, supported in part by the Institute for Arctic and Alpine Research (INSTAAR).
12 National Hurricane Center and Central Pacific Hurricane Center, Saffir-Simpson Hurricane Wind Scale. www.nhc.noaa.gov/aboutsshws.php
13 Richtel M (2023) Bomb Cyclone? Or Just Windy with a Chance of Hyperbole? When the Barometer Drops, the Volume of "Hyped Words" Rises, and Many Meteorologists Aren't Happy about It. *New York Times*, January 18.
14 The journal *Environment and History* is published by White Horse Press. See also: Strauss S and Orlove BS (2021) *Weather, Climate, Culture*. Routledge.
15 Jeffers R (2022) The Battle of Waterloo: Did the Weather Change History? #EveryWomanDreams blog. https://reginajeffers.blog/2022/06/13/the-battle-of-waterloo-did-the-weather-change-history/
16 Neumann J (1993) Great Historical Events That Were Significantly Affected by the Weather. Part 11: Meteorological Aspects of the Battle of Waterloo. *Bulletin of the American Meteorological Society* 74, 413–420.
17 Niderost E (2014) The Second Great War (and the Weather that Defined It) *WWII History*, 14 (1), 70–76.

18 Deunsing R (2021) How the Weather Impacted the Attack on Pearl Harbor 80 Years Ago. *CBS News*, December 7, www.cbs17.com/news/national-news/how-the-weather-impacted-the-attack-on-pearl-harbor-80-years-ago/

19 Neumann J (1975) Great Historical Events That Were Significantly Affected by the Weather: 1, The Mongol Invasions of Japan. *Bulletin of the American Meteorological Society* 56(11), 1167–1171.

20 Neumann J (1977) Great Historical Events That Were Significantly Affected by the Weather: 2, The Year Leading to the Revolution of 1789 in France. *Bulletin of the American Meteorological Society* 58(2), 163–168.

21 Neumann J (1978) Great Historical Events That Were Significantly Affected by the Weather: 3, The Cold Winter of 1657–58, the Swedish Army Crosses Denmark's Frozen Sea Areas. *Bulletin of the American Meteorological Society* 59(11), 1432–1437.

22 Neumann J and Lindgrén S (1979) Great Historical Events That Were Significantly Affected by the Weather: 4, The Great Famines in Finland and Estonia, 1695–97. *Bulletin of the American Meteorological Society* 60(7), 775–787.

23 Lindgrén S and Neumann J (1980) Great Historical Events That Were Significantly Affected by the Weather: 5, Some Meteorological Events of the Crimean War and Their Consequences. *Bulletin of the American Meteorological Society* 61(12), 1570–1583.

24 Bass J and Rankin Jr A (dir.) (1974) *The Year without a Santa Claus*. Rankin/Bass Productions.

25 Lemmin-Woolfrey, U (2024) This Country's Doctors Can Prescribe Four-week Spa Breaks to Frazzled Parents. *CNN*, August 1, www.cnn.com/travel/germany-kur-parents-spa-breaks-wellness/index.html

26 Howarth E and Hoffman MS (1984) A Multidimensional Approach to the Relationship between Mood and Weather. *British Journal of Psychology* 75(1), 15–23.

27 Keller MC, Fredrickson BL, Ybarra O, Côté S, Johnson K, Mikels J, Conway A, and Wager T (2005) A Warm Heart and a Clear Head: The Contingent Effects of Weather on Mood and Cognition. *Psychological Science* 16(9), 724–731.

28 Denissen JJ, Butalid L, Penke L, and Van Aken MA (2008) The Effects of Weather on Daily Mood: A Multilevel Approach. *Emotion* 8(5), 662.

29 Bosak-Schroeder C (2016) The Ecology of Health in Herodotus, Dicaearchus, and Agatharchides. In Kennedy RF and Jones-Lewis M (eds.), *The Routledge Handbook of Identity and the Environment in the Classical and Medieval Worlds.* Routledge, 29–44.

30 Hippocrates (1923) On Airs, Waters, and Places. In Adams F (trans.) *The Genuine Works of Hippocrates.* Sydenham Society, 190–222.

31 Avicenna (Abū 'Alī al-Ḥusayn ibn 'Abd Allāh ibn Sīnā) (1973) The Canon of Medicine (al-Qānūn fī'l-ṭibb). In Gruner CO and Shah MH (trans.), *Great Books of the Islamic World.* AMS Press, 185–186.

32 Lines 109–201 of *Works and Days*, in Evelyn-White HG (1920) *Hesiod, the Homeric Hymns, and Homerica.* W. Heinemann.

33 In Book III of *The Republic*, in Cooper JM and Hutchinson DS (1997) *Plato: Complete Works.* Hackett Publishing.

34 Malebranche N (2014) *Dialogues on Metaphysics.* Routledge; Leclerc GL and De Buffon C (2024) Natural History, General and Particular. In Hawley J (ed.), *Literature and Science, 1660–1834*, Part II, Vol. 5. Routledge, 219–244.

35 De Secondat Montesquieu C-L and baron de Montesquieu CdS (2005) *The Spirit of Laws.* The Lawbook Exchange, Ltd.

36 Immerwahr J (1992) Hume's Revised Racism. *Journal of the History of Ideas* 53(3), 481–486.
37 Hume D (1994) Of National Characters. In Haakonssen K (ed.), *Hume: Political Essays*. Cambridge University Press, 78–92, p. 90.
38 Klinke I and Bassin M (2018) Introduction: Lebensraum and Its Discontents. *Journal of Historical Geography* 61, 53–58.
39 Diamond JM (1998) *Guns, Germs and Steel: The Fates of Human Societies*. Random House.

Chapter 3

1 There's a very technical discussion here on dispositionalism regarding how we should understand our beliefs, our values, and their relation to the rest of the world. I don't think that discussion is particularly important to the upshot of this argument, so I'd like to sidestep it if possible. For more technical purposes, I think it's important to distinguish between propositional attitudes and dispositional attitudes, where the former are attitudes that are held "in mind" (i.e., as *propositions*) and dispositional attitudes are attitudes describing how we comport ourselves toward the world. Nevertheless, to remain somewhat agnostic on this discussion, I refer to dispositional *orientations*, though for clarity I also refer to these orientations as attitudes. Schwitzgebel E (2013) A Dispositional Approach to Attitudes: Thinking Outside of the Belief Box. In Nottelmann N (ed.), *New Essays on Belief: Constitution, Content and Structure*. Springer, 75–99.
2 Aurelius M (2023) *Meditations*. Reader's Library Classics, Book VIII, 47.
3 Among the competing anthropological theories explaining the rise of agriculture during the Neolithic period, there is the oasis theory, the hilly flanks hypothesis, the feasting model, the

propinquity theory, the evolution/intention theory, and the climate theory, to name just a few. Edwards PC (2020) The Beginnings of Agriculture. In Hollander D and Howe T (eds.), *A Companion to Ancient Agriculture*, 119–148.
4 Shkolnik A, Taylor CR, Finch V, and Borut A (1980) Why Do Bedouins Wear Black Robes in Hot Deserts? *Nature* 283 (5745), 373–375.
5 Grass NM (1955) *History of Hosiery: From the Piloi of Ancient Greece to the Nylons of Modern America*. New York: Fairchild Publications, Inc.
6 Evelyn-White HG (1920) *Hesiod, the Homeric hymns, and Homerica*. W. Heinemann, lines 540–560.
7 Colville A (2020) Yu the Great, Tamer of China's Greatest Floods. *The China Project*, August 24, https://thechinaproject.com/2020/08/24/yu-the-great-tamer-of-chinas-greatest-floods/
8 Gong X, Booker MD, Brown JA, Mann GC, and Gastfield AM (2019) Yu the Great Managed the Flood. In Aghasaleh R (ed.), *Children and Mother Nature*. Brill, 25–31.
9 EM-DAT (2023) *2022 Disasters in Numbers*. Centre for Research on the Epidemiology of Disasters.
10 Somerville J (1941) Umbrellaology, or, Methodology in Social Science. *Philosophy of Science* 8(4), 557–566.
11 Revkin A and Mechaley L (2018) *Weather: From Cloud Atlases to Climate Change*. Union Square+ ORM; Sangster W (1871) *Umbrellas and Their History*. Cassell, Petter, and Galpin.
12 Bernstein PL (1996) *Against the Gods: The Remarkable Story of Risk*. Wiley.
13 Brush CF (2016) Lighting Up the World. University of Michigan Electrical and Computer Engineering Department, April 12, https://ece.engin.umich.edu/stories/charles-f-brush
14 Office of Energy Efficiency and Renewable Energy. US Department of Energy (2024) How Much Power is 1

Gigawatt? August 21, www.energy.gov/eere/articles/how-much-power-1-gigawatt; Ward C (2024) How Much Power is 1.21 Gigawatts, Anyway? The Science Behind Back to the Future. March 19, www.syfy.com/syfy-wire/what-is-121-gigawatts-anyway-the-science-behind-back-to-the-future; Glassie J (2007) Lightning Farms. *New York Times Magazine*, December 9, www.nytimes.com/2007/12/09/magazine/09lightningfarm.html?_r=1

15 Stanley-Becker I, Partlow J and Sanchez YW (2023) How a Saudi Firm Tapped a Gusher of Water in Drought-stricken Arizona. *Washington Post*, July 16, www.washingtonpost.com/politics/2023/07/16/fondomonte-arizona-drought-saudi-farm-water/

16 Costanza R, d'Arge R, De Groot R, Farber S, Grasso M, Hannon B, Limburg K, Naeem S, O'Neill RV, and Paruelo J (1998) The Value of Ecosystem Services: Putting the Issues in Perspective. *Ecological Economics* 25(1), 67–72; Costanza R, De Groot R, Braat L, Kubiszewski I, Fioramonti L, Sutton P, Farber S, and Grasso M (2017) Twenty Years of Ecosystem Services: How Far Have We Come and How Far Do We Still Need to Go? *Ecosystem Services* 28, 1–16; Daily GC (1997) Introduction: What Are Ecosystem Services. *Nature's Services: Societal Dependence on Natural Ecosystems* 1(1), 1–10, https://doi.org/10.12987/9780300188479-039; Daily GC (2013) Nature's Services: Societal Dependence on Natural Ecosystems (1997). In Robin L, Sörlin S, and Warde P (eds.), *The Future of Nature*. Yale University Press, 454–464.

17 Ecosystem Services. *Climate Hubs: US Department of Agriculture.* www.climatehubs.usda.gov/ecosystem-services

18 Reid WV et al (2005) Ecosystems and Human Well-Being: Synthesis. *Millenium Ecosystem Assessment*. Island Press.

19 UN Environment Programme. The Economics of Ecosystems and Biodiversity (TEEB). www.unep.org/topics/teeb

20 Schröter M, Van der Zanden EH, van Oudenhoven AP, Remme RP, Serna-Chavez HM, De Groot RS, and Opdam P (2014) Ecosystem Services as a Contested Concept: A Synthesis of Critique and Counter-Arguments. *Conservation Letters* 7(6), 514–523.
21 Badorf F and Hoberg K (2020) The Impact of Daily Weather on Retail Sales: An Empirical Study in Brick-and-Mortar Stores. *Journal of Retailing and Consumer Services* 52, 101921, https://doi.org/10.1016/j.jretconser.2019.101921
22 Heidegger M (1977) *The Question concerning Technology*. Harper and Row.
23 Beck U (1992) *Risk Society: Towards a New Modernity*. Sage; Beck U (2012) World Risk Society. In Friis JKBO, Pedersen SA, and Hendricks VF (eds.), *A Companion to the Philosophy of Technology*. Wiley-Blackwell, 495–499; Giddens A (1990) *The Consequences of Modernity*. Stanford University Press.
24 Bernstein PL (1996) *Against the Gods: The Remarkable Story of Risk*. Wiley.

Chapter 4

1 Goodrich worked for KVOA-TV Channel 4 in Tucson from 1977 until his retirement in 1998.
2 Ramis H (dir.) (1993) *Groundhog Day*. Columbia Pictures.
3 I acknowledge that "weathermen" is a gendered, and in some respects dated, term, but I think it's also somewhat of a dated idea that weathermen are empty talking heads. The news meteorologists of roughly the past thirty years have not only been much more diverse but also considerably more skilled in consulting multiple sources. Arguably the introduction of the Weather Channel in 1982 – which began as a pay television channel and focused exclusively on the weather – brought both increased entertainment value to the reporting of the weather

and much more sophisticated meteorological science to the public.
4 Orlove BS, Chiang JC, and Cane MA (2002) Ethnoclimatology in the Andes: A Cross-disciplinary Study Uncovers a Scientific Basis for the Scheme Andean Potato Farmers Traditionally Use to Predict the Coming Rains. *American Scientist* 90(5), 428–435.
5 Bartlett AK (1909) The Wet and Dry Moon. *Popular Astronomy* 17, 11–14.
6 Ornan T (2001) The Bull and Its Two Masters: Moon and Storm Deities in Relation to the Bull in Ancient Near Eastern art. *Israel Exploration Journal* 51(1), 1–26.
7 Boyer CB (1958) The Tertiary Rainbow: An Historical Account. *Isis* 49(2), 141–154.
8 Aristotle, *Politics*, 1259a, www.perseus.tufts.edu/hopper/text?doc = Perseus%3Atext%3A1999.01.0058%3Abook%3D1%3Asection%3D1259a
9 Lloyd GER (1964) The Hot and the Cold, the Dry and the Wet in Greek Philosophy. *The Journal of Hellenic Studies* 84, 92–106.
10 "Nature abhors a vacuum, and if I can only walk with sufficient carelessness I am sure to be fulfilled." Thoreau HD (1881) *Early Spring in Massachusetts: From the Journal of Henry D. Thoreau*. Houghton, Mifflin and Company, 34–35.
11 "Nature abhors a vacuum: whenever people do not know the truth, they fill the gaps with conjecture."
12 Brunschön CW and Sider D (2007) *Theophrastus of Eresus: On Weather Signs*. Brill.
13 Bostock J and Riley HT (1855) *The Natural History of Pliny*. HG Bohn, book II, part III, chapter 46.
14 Ahmed A, Philippe P, and Mohamed I (2000) Eco-ethological Data According to Ǧāḥiẓ through His Work Kitāb al-Ḥayawān (The Book of Animals). *Arabica: Revue d' Etudes Arabes* 47(2), 278–286.

15 Adamson P (2006) *Al-Kindī*. Oxford University Press, 181–248.
16 Gunther RT (1929) *Chaucer and Messahalla on the Astrolabe*, Vol. 5. Oxford University Press.
17 Bacon F (2011) Preparative toward Natural and Experimental History. In Ellis RL, Heath DD, and Spedding J (eds.), *The Works of Francis Bacon*, Vol. 4, *Translations of the Philosophical Works 1*. Cambridge University Press, 249–250.
18 Kepler J (2010) *The Six-Cornered Snowflake*. Paul Dry Books.
19 Descartes, R (2001) *Discourse on Method, Optics, Geometry, and Meteorology*, trans. Olscamp PJ. Hackett, 263.
20 Brissey P (2012) Descartes and the Meteorology of the World. *Societate si politica* 6(2), 99–113.
21 Frisinger HH (1983) The Barometer. In Frisinger HH (ed.), *The History of Meteorology: To 1800*. American Meteorological Society, 66–79.
22 Golinski J (2021) Time, Talk, and the Weather in Eighteenth-Century Britain. In Strauss S and Orlove BS (eds.), *Weather, Climate, Culture*. Routledge, 17–38.
23 Franklin B (1784) Meteorological Imaginations and Conjectures. *Manchester Literary and Philosophical Society Memoirs and Proceedings* 2, 122.
24 Rorsch A (2007) Climate Science and the Phlogiston Theory: Weighing the Evidence. *Energy & Environment* 18(3/4), 441–447.
25 Kuhn T (1970) *The Structure of Scientific Revolutions*, 2nd edition. University of Chicago Press.
26 Jefferson T (1998) *Notes on the State of Virginia*. Penguin.
27 Fleming JR (1998) *Historical Perspectives on Climate Change*. Oxford University Press.
28 The best compilation of this history that I have been able to find, and that I was fortunate to stumble upon, seems oddly related to an expansive stamp collecting ("thematic philately")

effort to catalogue all weather-related stamps, otherwise known as "meteorophilately." https://rammb.cira.colostate.edu/dev/hillger/meteorologist.htm

29 "When it is evening, you say, 'It will be fair weather; for the sky is red.' And in the morning, 'It will be stormy today, for the sky is red and threatening.'" (Matthew 6:2–3). In US Library of Congress (2024) Is the Old Adage "Red Sky at Night, Sailor's Delight. Red Sky in Morning, Sailor's Warning" True, or Is It Just an Old Wives' Tale? Science Reference Section, June 4, www.loc.gov/everyday-mysteries/meteorology-climatology/item/is-the-old-adage-red-sky-at-night-sailors-delight-red-sky-in-morning-sailors-warning-true-or-is-it-just-an-old-wives-tale/

30 (1920) A Conspiracy to Keep Us Ignorant. *Bulletin of the American Meteorological Society* 1(9), 102a–102; Baldwin HI (1920) Errors of House Thermometers. Bulletin of the American Meteorological Society 1(4), 39a–39.

31 Redway JW (1920) Do Climates Change? *Bulletin of the American Meteorological Society* 1(1), 86–87.

32 Harper KC (2012) *Weather by the Numbers: The Genesis of Modern Meteorology.* MIT Press.

33 Gooley T (2021) *The Secret World of Weather: How to Read Signs in Every Cloud, Breeze, Hill, Street, Plant, Animal, and Dewdrop.* The Experiment.

34 Hardin R (2003) *Indeterminacy and Society.* Princeton University Press; Wynne B (1992) Uncertainty and Environmental Learning: Reconceiving Science and Policy in the Preventive Paradigm. *Global Environmental Change* 2(2), 111–127; Hale B and Shockley K (2023) Risk Mismanagement: The Illusion of Control in Indeterminate Systems. In Placani A and Broadhead S (eds.), *Risk and Responsibility in Context.* Routledge, 51–70.

35 According to some sources, the term "Eskimo" is derogatory and should be retired from use. However, the cliché seems

bound up in the caricature of Inuit people as Eskimo. Robson D (2012) Are There Really 50 Eskimo Words for Snow? *New Scientist*, December 18.
36 Bush K (2011) 50 Words for Snow. Fish People.
37 Plato (1995) *Phaedrus*. Hackett.
38 Peters M (2006) Not Your Father's Cliché. *Columbia Journalism Review* 45(2), 14–15.
39 Jacobson SA (1984) *Yup'ik Eskimo Dictionary*. ERIC.
40 *Radiolab* (2012) Colors. May 12, https://radiolab.org/podcast/211119-colors
41 Chang H (2004) *Inventing Temperature: Measurement and Scientific Progress*. Oxford University Press; McCaskey JP (2020) History of "Temperature": Maturation of a Measurement Concept. *Annals of Science* 77(4), 399–444; Dickinson J (2025) Temperature Changes: The Conceptual Realignment of a Quantity Term. *Studies in History and Philosophy of Science* 109, 47–57.
42 De Spinoza B (2012) Spinoza's Algebraic Calculation of the Rainbow and Calculation of Chances. Edited and translated with an introduction, explanatory notes and appendix by Petry MJ. Springer Science & Business Media.
43 Bernstein PL (1996) *Against the Gods: The Remarkable Story of Risk*. Wiley.
44 Ibid. In a somewhat clever gag, Bernstein enumerates the pages of his discussion of the invention of numbers in the second chapter all in Roman numerals. I have not been able to track down the original source for this Whitehead quote, though it is present in many places.

Chapter 5

1 Reiley L (2024) Cutting-Edge Tech Made This Tiny Country a Major Exporter of Food. *Washington Post*, November 21.

2 Seton JM and Seton ET (1930) *The Rhythm of the Redman: In Song, Dance and Decoration*. Ronald Press Co.; Parkman EB (1993) Creating Thunder: The Western Rain-Making Process. *Journal of California and Great Basin Anthropology* 15(1), 90–110.
3 Following a convention recommended by the US-based National Museum of the American Indian, I am using the term "Indian" here to refer to tribes of first peoples and indigenous populations, mostly in North America. I acknowledge, of course, that there is no single American Indian culture or language and that different cultural groups may prefer a different designation. National Museum of the American Indian (2025). The Impact of Words and Tips for Using Appropriate Terminology: Am I Using the Right Word? https://americanindian.si.edu/nk360/informational/impact-words-tips
4 Parkman EB (1993) Creating Thunder: The Western Rain-Making Process. *Journal of California and Great Basin Anthropology* 15(1), 90–110.
5 Hitch S (2015) Sacrifice. In Wilkins J and Nadeau R (eds.), *A Companion to Food in the Ancient World*, 335–347.
6 Homer (1919) *The Odyssey*. GP Putnam's Sons.
7 Benjamin T (2009) *The Atlantic World: Europeans, Africans, Indians and Their Shared History, 1400–1900*. Cambridge University Press.
8 Stemp WJ (2016) Explorations in Ancient Maya Blood-Letting: Experimentation and Microscopic Use-Wear Analysis of Obsidian Blades. *Journal of Archaeological Science: Reports* 7, 368–378.
9 Guarino B (2019) History's Largest Child Sacrifice Was a Response to Devastating Weather, Archeologists Say. *Washington Post*, March 6, www.washingtonpost.com/science/2019/03/06/historys-largest-child-sacrifice-was-response-devastating-weather-archaeologists-say; Stewart-Kroeker S

(2022) Sacrifice in Environmental Ethics and Theology. *The Journal of Religion* 102(2), 237-261.
10 Iannone G (2014) *The Great Maya Droughts in Cultural Context: Case Studies in Resilience and Vulnerability.* University Press of Colorado; Gill RB (2000) *The Great Maya Droughts: Water, Life, and Death.* UNM Press; Fagan B (2009) *Floods, Famines, and Emperors: El Niño and the Fate of Civilizations.* Basic Books.
11 See, for instance, Plato's *Timeus*: Cooper JM and Hutchinson DS (1997) *Plato: Complete Works.* Hackett.
12 St. Augustine was a bit more of a Platonist than an Aristotelian, though many after Aquinas, 900 years later, interpreted his position through an Aristotelian lens. In any case, both Plato and Aristotle held teleological positions, with the former's teleology being more supernaturalistic and the latter's more naturalistic. During the Middle Ages, both teleological positions were imbued with monotheism.
13 White AD (1896) *A History of the Warfare of Science with Theology in Christendom.* George Braziller.
14 Oster E (2004) Witchcraft, Weather and Economic Growth in Renaissance Europe. *Journal of Economic Perspectives* 18(1), 215-228; Fagan B (2019) *The Little Ice Age: How Climate Made History 1300-1850.* Hachette.
15 Mackay CS (2009) *The Hammer of Witches: A Complete Translation of the Malleus Maleficarum.* Cambridge University Press.
16 Oster E (2004) Witchcraft, Weather and Economic Growth in Renaissance Europe. *Journal of Economic Perspectives* 18(1), 215-228.
17 Ember CR, Skoggard I, Felzer B, Pitek E, and Jiang M (2021) Climate Variability, Drought, and the Belief that High Gods Are Associated with Weather in Nonindustrial Societies. *Weather, Climate, and Society* 13(2), 259-272.

18 Golinski J (2021) Time, Talk, and the Weather in Eighteenth-Century Britain. In Strauss S and Orlove B (eds.), *Weather, Climate, Culture*. Routledge, 17–38.
19 Hardwick J and Stephens RJ (2020) Acts of God: Continuities and Change in Christian Responses to Extreme Weather Events from Early Modernity to the Present. *Wiley Interdisciplinary Reviews: Climate Change* 11(2), e631, https://doi.org/10.1002/wcc.631
20 Coverley DM (2017) *Global Warming or God's Warning: A Prophetic Explanation for the Strange and Unusual Events in the Skies, on the Land, in the Waters, and with the Weather*. Westbow Press; Shao W (2017) Weather, Climate, Politics, or God? Determinants of American Public Opinions toward Global Warming. *Environmental Politics* 26(1), 71–96.
21 Martin C (2010) The Ends of Weather: Teleology in Renaissance Meteorology. *Journal of the History of Philosophy* 48(3), 259–282.
22 Bacon F (2008) New Atlantis [1627]. In Bruce S (ed.), *Three Early Modern Utopias: Utopia, New Atlantis, The Isle of Pines*, Oxford University Press, 152–155.
23 Mechanistic versus organic models are taken up in some detail in another book in the series. See, for instance, Ruse M (2021) *A Philosopher Looks at Human Beings*. Cambridge University Press, 21–47.
24 Spence CC (1961) A Brief History of Pluviculture. *The Pacific Northwest Quarterly* 52(4), 129–138.
25 Further digging suggests that Hatfield (or perhaps one of his relatives) had been at it for quite a long time, since at least as long as 1904. See, for instance, Guinn JM (1904) Rain and Rainmakers. *Annual Publication of the Historical Society of Southern California and of the Pioneers of Los Angeles County* 6 (2), 171–176; Vercler HR (1920) More about the "Rain-Maker." *Bulletin of the American Meteorological Society* 1(7–8), 80–82.

26 Jordan DS (1925) The Art of Pluviculture. *Science* 62(1595), 81–82.
27 Ibid.
28 Holmes TT (2016) Hail Cannons, the Devices That Supposedly Blast Away Bad Weather. *Atlas Obscura*, March 29, www.atlasobscura.com/articles/hail-cannons-the-devices-that-supposedly-blast-away-bad-weather
29 Merali Z (2005) Oil on Troubled Waters May Stop Hurricanes. *New Scientist*, July 25, www.newscientist.com/article/dn7726-oil-on-troubled-waters-may-stop-hurricanes/
30 National Oceanic & Atmospheric Administration. Tropical Cyclones Frequently Asked Questions. www.aoml.noaa.gov/hrd/tcfaq/C5d.html
31 Willoughby H, Jorgensen D, Black R, and Rosenthal S (1985) Project STORMFURY: A Scientific Chronicle 1962–1983. *Bulletin of the American Meteorological Society* 66(5), 505–514.
32 Gentry RC (1970) Hurricane Debbie Modification Experiments, August 1969. *Science* 168(3930), 473–475.
33 Dennett DC (1989) *The Intentional Stance*. MIT Press.
34 McElwee P (2020) The Origins of Ecocide: Revisiting the Ho Chi Minh Trail in the Vietnam War. *Arcadia* Spring (20), https://doi.org/10.5282/rcc/9039; Hambling D (2023) "Make Mud, Not War": How U.S. Used Weather Warfare in Vietnam. *The Guardian*, August 10, www.theguardian.com/news/2023/aug/10/make-mud-not-war-us-weather-warfare-vietnam-commando-lava-ho-chi-minh
35 United Nations (1976) Convention on the Prohibition of Military or Any Other hostile Use of Environmental Modification Techniques. 1108 UNTS 151, December 10, https://treaties.un.org/Pages/ViewDetails.aspx?src=IND&mtdsg_no=XXVI-1&chapter=26&clang=_en
36 Rose DC, Bhattacharya M, de Boon A, Dhulipala RK, Price C, and Schillings J (2022) The Fourth Agricultural Revolution:

Technological Developments in Primary Food Production. In Sage CL (ed.), *A Research Agenda for Food Systems*. Edward Elgar, 151–174; Rose DC, Barkemeyer A, De Boon A, Price C, and Roche D (2023) The Old, the New, or the Old Made New? Everyday Counter-narratives of the so-called Fourth Agricultural Revolution. *Agriculture and Human Values* 40(2), 423–439.

BIBLIOGRAPHY

Adamson P (2006) *Al-Kindī*. Oxford University Press.

Ahmed A, Philippe P, and Mohamed I (2000) Eco-ethological Data according to Ǧāḥiẓ through His Work Kitāb al-Ḥayawān (The Book of Animals). *Arabica: Revue d' Etudes Arabes* 47(2), 278–286.

American Meteorological Society (1920) A Conspiracy to Keep Us Ignorant. *Bulletin of the American Meteorological Society* 1(9), 102a–102.

Aurelius M (2023) *Meditations*. Reader's Library Classics.

Bacon F (2008) New Atlantis [1627]. In Bruce S (ed.), *Three Early Modern Utopias: Utopia, New Atlantis, The Isle of Pines*, Oxford University Press, 152–155.

Bacon F (2011) Preparative toward Natural and Experimental History. In Ellis RL, Heath DD, and Spedding J (eds.), *The Works of Francis Bacon, Vol. 4, Translations of the Philosophical Works 1*. Cambridge University Press, 249–250.

Badorf F and Hoberg K (2020) The Impact of Daily Weather on Retail Sales: An Empirical Study in Brick-and-Mortar Stores. *Journal of Retailing and Consumer Services* 52, 101921, https://doi.org/10.1016/j.jretconser.2019.101921.

Baldwin HI (1920) Errors of House Thermometers. *Bulletin of the American Meteorological Society* 1(4), 39a–39.

Bartlett AK (1909) The Wet and Dry Moon. *Popular Astronomy* 17, 11–14.

Bass J and Rankin Jr A (dir.) (1974) *The Year without a Santa Claus*. Rankin/Bass Productions.

Beck U (1992) *Risk Society: Towards a New Modernity*. Sage.

Beck U (2012) World Risk Society. In Friis JKBO, Pedersen SA, and Hendricks VF (eds.), *A Companion to the Philosophy of Technology*. Wiley-Blackwell. 495–499.

Benjamin T (2009) *The Atlantic World: Europeans, Africans, Indians and Their Shared History, 1400–1900*. Cambridge University Press.

Bernstein PL (1996) *Against the Gods: The Remarkable Story of Risk*. Wiley.

Bosak-Schroeder C (2016) The Ecology of Health in Herodotus, Dicaearchus, and Agatharchides. In Kennedy RF and Jones-Lewis M (eds.), *The Routledge Handbook of Identity and the Environment in the Classical and Medieval Worlds*. Routledge, 29–44.

Bostock J and Riley HT (1855) *The Natural History of Pliny*. HG Bohn.

Boyer CB (1958) The Tertiary Rainbow: An Historical Account. *Isis* 49(2), 141–154.

Brissey P (2012) Descartes and the Meteorology of the World. *Societate si politica* 6(2), 99–113.

Brunschön CW and Sider D (2007) *Theophrastus of Eresus: On Weather Signs*. Brill.

Bulwer-Lytton E (1833) *Paul Clifford*. Baudry's European Library.

Chang H (2004) *Inventing Temperature: Measurement and Scientific Progress*. Oxford University Press.

Chang LW, Kirgios EL, Mullainathan S, and Milkman KL (2024) Does Counting Change What Counts? Quantification Fixation Biases Decision-Making. *Proceedings of the National Academy of Sciences* 121(46), e2400215121, https://doi.org/doi:10.1073/pnas.2400215121.

Clara B-S (2015) The Ecology of Health in Herodotus, Dicaearchus, and Agatharchides. In Kennedy RF and Jones-Lewis M (eds.), *The Routledge Handbook of Identity and the Environment in the Classical and Medieval Worlds*. Routledge, 29–44.

Cooper JM and Hutchinson DS (1997) *Plato: Complete Works*. Hackett.

Costanza R, d'Arge R, De Groot R, Farber S, Grasso M, Hannon B, Limburg K, Naeem S, O'Neill RV, and Paruelo J (1998) The Value of Ecosystem Services: Putting the Issues in Perspective. *Ecological Economics* 25(1), 67–72.

Costanza R, De Groot R, Braat L, Kubiszewski I, Fioramonti L, Sutton P, Farber S, and Grasso M (2017) Twenty Years of Ecosystem Services: How Far Have We Come and How Far Do We Still Need to Go? *Ecosystem Services* 28, 1–16.

Coverley DM (2017) *Global Warming or God's Warning: A Prophetic Explanation for the Strange and Unusual Events in the Skies, on the Land, in the Waters, and with the Weather*. Westbow Press.

Daily GC (1997) Introduction: What Are Ecosystem Services? *Nature's Services: Societal Dependence on Natural Ecosystems* 1(1), 1–10, https://doi.org/10.12987/9780300188479-039.

Daily GC (2013) Nature's Services: Societal Dependence on Natural Ecosystems [1997]. In Robin L, Sörlin S, and Warde P (eds.), *The Future of Nature*. Yale University Press, 454–464.

De Secondat Montesquieu C-L and Baron de Montesquieu CdS (2005) *The Spirit of Laws*. The Lawbook Exchange, Ltd.

De Spinoza B (2012) *Spinoza's Algebraic Calculation of the Rainbow and Calculation of Chances: Edited and Translated with an Introduction, Explanatory Notes and an Appendix by Michael J. Petry*. Springer Science & Business Media.

Denissen JJ, Butalid L, Penke L, and Van Aken MA (2008) The Effects of Weather on Daily Mood: A Multilevel Approach. *Emotion* 8(5), 662.

Dennett DC (1989) *The Intentional Stance*. MIT Press.

Dennis J (2013) *It's Raining Frogs and Fishes: Four Seasons of Natural Phenomena and Oddities of the Sky*. Diversion Books.

Diamond JM (1998) *Guns, Germs and Steel: The Fates of Human Societies*. Random House.

Dickinson J (2025) Temperature Changes: The Conceptual Realignment of a Quantity Term. *Studies in History and Philosophy of Science* 109, 47–57.

Edwards PC (2020) The Beginnings of Agriculture. In Hollander D and Howe T (eds.), *A Companion to Ancient Agriculture*, 119–148.

EM-DAT (2023) 2022 Disasters in Numbers. Centre for Research on the Epidemiology of Disasters (CRED).

Ember CR, Skoggard I, Felzer B, Pitek E, and Jiang M (2021) Climate Variability, Drought, and the Belief that High Gods Are Associated with Weather in Nonindustrial Societies. *Weather, Climate, and Society* 13(2), 259–272.

Evelyn-White HG (1920) *Hesiod, the Homeric hymns, and Homerica*. W. Heinemann.

Fagan B (2009) *Floods, Famines, and Emperors: El Niño and the Fate of Civilizations*. Basic Books.

Fagan B (2019) *The Little Ice Age: How Climate Made History 1300–1850*. Hachette.

Fleming JR (1998) *Historical Perspectives on Climate Change*. Oxford University Press.

Franklin B (1784) Meteorological Imaginations and Conjectures. *Manchester Literary and Philosophical Society Memoirs and Proceedings* 2, 122.

Frisinger HH (1983) The Barometer. In Frisinger HH (ed.), *The History of Meteorology: To 1800*. American Meteorological Society, 66–79.

Gentry RC (1970) Hurricane Debbie Modification Experiments, August 1969. *Science* 168(3930), 473–475.

Giddens A (1990) *The Consequences of Modernity*. Stanford University Press.

Gill RB (2000) *The Great Maya Droughts: Water, Life, and Death*. UNM Press.

Glassie J (2007) Lightning Farms. *New York Times Magazine*, December 9, www.nytimes.com/2007/12/09/magazine/09lightningfarm.html?_r=1.

Golinski J (2021) Time, Talk, and the Weather in Eighteenth-Century Britain 1. In Strauss S and Orlove B (eds.), *Weather, Climate, Culture*. Routledge, 17–38.

Gong X, Booker MD, Brown JA, Mann GC, and Gastfield AM (2019) Yu the Great Managed the Flood. In Aghasaleh R (ed.), *Children and Mother Nature*. Brill, 25–31.

Gooley T (2021) *The Secret World of Weather: How to Read Signs in Every Cloud, Breeze, Hill, Street, Plant, Animal, and Dewdrop*. The Experiment.

Grass NM (1955) *History of Hosiery: From the Piloi of Ancient Greece to the Nylons of Modern America*. Fairchild Publications, Inc.

Grzega J (2022) Climatic Conditions and Lexis: Some Diachronic Notes on Weather-Related Words in English and Other European Languages. *Transactions of the Philological Society* 120(2), 320–331.

Guarino B (2019) History's Largest Child Sacrifice Was a Response to Devastating Weather, Archeologists Say. *Washington Post*, www.washingtonpost.com/science/2019/03/06/historys-largest-child-sacrifice-was-response-devastating-weather-archaeologists-say/.

Guinn JM (1904) Rain and Rainmakers. *Annual Publication of the Historical Society of Southern California and of the Pioneers of Los Angeles County* 6(2), 171–176.

Gunther RT (1929) *Chaucer and Messahalla on the Astrolabe*, Vol. 5. Oxford University Press.

Hale B and Shockley K (2023) Risk Mismanagement: The Illusion of Control in Indeterminate Systems. In Placani A and

Broadhead S (eds.), *Risk and Responsibility in Context.* Routledge, 51–70.

Hambling D (2023) "Make Mud, Not War": How U.S. Used Weather Warfare in Vietnam. *The Guardian*, August 10, www.theguardian.com/news/2023/aug/10/make-mud-not-war-us-weather-warfare-vietnam-commando-lava-ho-chi-minh.

Hardin R (2003) *Indeterminacy and Society.* Princeton University Press.

Hardwick J and Stephens RJ (2020) Acts of God: Continuities and Change in Christian Responses to Extreme Weather Events from Early Modernity to the Present. *Wiley Interdisciplinary Reviews: Climate Change* 11(2), e631, https://doi.org/10.1002/wcc.631.

Harper KC (2012) *Weather by the Numbers: The Genesis of Modern Meteorology.* MIT Press.

Heidegger M (1977) *The Question concerning Technology.* New York: Harper and Row.

Hitch S (2015) Sacrifice. In Wilkins J and Nadeau R (eds.), *A Companion to Food in the Ancient World*, 335–347.

Holmes TT (2016) Hail Cannons, the Devices That Supposedly Blast Away Bad Weather. *Atlas Obscura*, March 29, www.atlasobscura.com/articles/hail-cannons-the-devices-that-supposedly-blast-away-bad-weather.

Homer (1919) *The Odyssey.* GP Putnam's Sons.

Howarth E and Hoffman MS (1984) A Multidimensional Approach to the Relationship between Mood and Weather. *British Journal of Psychology* 75(1), 15–23.

Hume D (1994) Of National Characters. In Hume D and Haakonssen K (eds.), *Hume: Political Essays.* Cambridge University Press, 78–92.

Iannone G (2014) *The Great Maya Droughts in Cultural Context: Case Studies in Resilience and Vulnerability.* University Press of Colorado.

Immerwahr J (1992) Hume's Revised Racism. *Journal of the History of Ideas* 53(3), 481–486.

Izon G, Zerkle AL, Williford KH, Farquhar J, Poulton SW, and Claire MW (2017) Biological Regulation of Atmospheric Chemistry en Route to Planetary Oxygenation. *Proceedings of the National Academy of Sciences* 114(13), E2571–E2579.

Jacobson SA (1984) *Yup'ik Eskimo Dictionary*. ERIC.

Jefferson T (1998) *Notes on the State of Virginia*. Penguin.

Jordan DS (1925) The Art of Pluviculture. *Science* 62(1595), 81–82.

Juliano TW, Lareau N, Frediani ME, Shamsaei K, Eghdami M, Kosiba K, Wurman J, DeCastro A, Kosović B, and Ebrahimian H (2023) Toward a Better Understanding of Wildfire Behavior in the Wildland–Urban Interface: A Case Study of the 2021 Marshall Fire. *Geophysical Research Letters* 50(10), e2022GL101557, https://doi.org/10.1029/2022GL101557.

Juster N (2011) *The Phantom Tollbooth*, 50th anniversary edition. Knopf Books for Young Readers.

Keller MC, Fredrickson BL, Ybarra O, Côté S, Johnson K, Mikels J, Conway A, and Wager T (2005) A Warm Heart and a Clear Head: The Contingent Effects of Weather on Mood and Cognition. *Psychological Science* 16(9), 724–731.

Kennedy RF and Jones-Lewis M (eds). (2016) *The Routledge Handbook of Identity and the Environment in the Classical and Medieval Worlds*. Routledge.

Kepler J (2010) *The Six-Cornered Snowflake*. Paul Dry Books.

Klinke I and Bassin M (2018) Introduction: Lebensraum and Its Discontents. *Journal of Historical Geography* 61, 53–58.

Knox B (1997) *The Odyssey*. Penguin.

Koppe C, Kovats S, Jendritzky G, and Menne B (2004) *Heat-Waves: Risks and Responses*. World Health Organization, Regional Office for Europe.

Kuhn T (1970) *The Structure of Scientific Revolutions*, 2nd edition. University of Chicago Press.

Leclerc GL and De Buffon C (2024) Natural History, General and Particular. In Hawley J (ed.), *Literature and Science, 1660–1834*, Part II, Vol. 5. Routledge, 219–244.

Lindgrén S and Neumann J (1980) Great Historical Events that Were Significantly Affected by the Weather: 5, Some Meteorological Events of the Crimean War and Their Consequences. *Bulletin of the American Meteorological Society* 61(12), 1570–1583.

Lloyd GER (1964) The Hot and the Cold, the Dry and the Wet in Greek Philosophy. *The Journal of Hellenic Studies* 84, 92–106.

Macauley D (2010) *Elemental Philosophy: Earth, Air, Fire, and Water as Environmental Ideas*. State University of New York Press.

Mackay CS (2009) *The Hammer of Witches: A Complete Translation of the Malleus Maleficarum*. Cambridge University Press.

Malebranche N (2014) *Dialogues on Metaphysics*. Routledge.

Martin C (2010) The Ends of Weather: Teleology in Renaissance Meteorology. *Journal of the History of Philosophy* 48(3), 259–282.

McCaskey JP (2020) History of "Temperature": Maturation of a Measurement Concept. *Annals of Science* 77(4), 399–444.

McElwee P (2020) The Origins of Ecocide: Revisiting the Ho Chi Minh Trail in the Vietnam War. *Arcadia Spring* (20), https://doi.org/10.5282/rcc/9039.

McKim K (2013) *Cinema as Weather: Stylistic Screens and Atmospheric Change*. Routledge.

Merali Z (2005) Oil on Troubled Waters May Stop Hurricanes. *New Scientist*, July 25, www.newscientist.com/article/dn7726-oil-on-troubled-waters-may-stop-hurricanes/.

Neumann J (1975) Great Historical Events That Were Significantly Affected by the Weather: 1. The Mongol Invasions of Japan. *Bulletin of the American Meteorological Society* 56(11), 1167–1171.

Neumann J (1977) Great Historical Events That Were Significantly Affected by the Weather: 2, The Year Leading to the Revolution of 1789 in France. *Bulletin of the American Meteorological Society* 58(2), 163–168.

Neumann J (1978) Great Historical Events That Were Significantly Affected by the Weather: 3, The Cold Winter of 1657–58, the Swedish Army Crosses Denmark's Frozen Sea Areas. *Bulletin of the American Meteorological Society* 59(11), 1432–1437.

Neumann J and Lindgrén S (1979) Great Historical Events That Were Significantly Affected by the Weather: 4, The Great Famines in Finland and Estonia, 1695–97. *Bulletin of the American Meteorological Society* 60(7), 775–787.

Orlove BS, Chiang JC, and Cane MA (2002) Ethnoclimatology in the Andes: A Cross-disciplinary Study Uncovers a Scientific Basis for the Scheme Andean Potato Farmers Traditionally Use to Predict the Coming Rains. *American Scientist* 90(5), 428–435.

Ornan T (2001) The Bull and Its Two Masters: Moon and Storm Deities in Relation to the Bull in Ancient Near Eastern Art. *Israel Exploration Journal* 51(1), 1–26.

Oster E (2004) Witchcraft, Weather and Economic Growth in Renaissance Europe. *Journal of Economic Perspectives* 18(1), 215–228.

Parkman EB (1993) Creating Thunder: The Western Rain-Making Process. *Journal of California and Great Basin Anthropology* 15(1), 90–110.

Parrish SS (2012) Faulkner and the Outer Weather of 1927. *American Literary History* 24(1), 34–58.

Peters M (2006) Not Your Father's Cliché. *Columbia Journalism Review* 45(2), 14–15.

Plato (1995) *Phaedrus*. Hackett.

Putnam H (1973) Meaning and Reference. *The Journal of Philosophy* 70(19), 699–711.

Ramis H (dir.) (1993) *Groundhog Day*. Columbia Pictures.

Reiley L (2024) Cutting-Edge Tech Made This Tiny Country a Major Exporter of Food. *Washington Post*, November 21.

Revkin A and Mechaley L (2018) *Weather: From Cloud Atlases to Climate Change*. Union Square+ ORM.

Richtel M (2023) Bomb Cyclone? Or Just Windy with a Chance of Hyperbole? When the Barometer Drops, the Volume of "Hyped Words" Rises, and Many Meteorologists Aren't Happy about It. *New York Times*, January 18.

Robson D (2012) Are There Really 50 Eskimo Words for Snow? *New Scientist*, www.newscientist.com/article/mg21628962-800-are-there-really-50-eskimo-words-for-snow/.

Rorsch A (2007) Climate Science and the Phlogiston Theory: Weighing the Evidence. *Energy & Environment* 18(3/4), 441–447.

Rose DC, Barkemeyer A, De Boon A, Price C, and Roche D (2023) The Old, the New, or the Old Made New? Everyday Counter-narratives of the so-called Fourth Agricultural Revolution. *Agriculture and Human Values* 40(2), 423–439.

Rose DC, Bhattacharya M, de Boon A, Dhulipala RK, Price C, and Schillings J (2022) The Fourth Agricultural Revolution: Technological Developments in Primary Food Production. In Sage CL (ed.), *A Research Agenda for Food Systems*. Edward Elgar, 151–174.

Ruse M (2021) *A Philosopher Looks at Human Beings*. Cambridge University Press.

Russell BW (2016) All the Gods of the World: Modern Maya Ritual in Yucatán, Mexico. *Ethnoarchaeology* 8(1), 4–29.

Sagan C and Mullen G (1972) Earth and Mars: Evolution of Atmospheres and Surface Temperatures. *Science* 177(4043), 52–56.

Sangster W (1871) *Umbrellas and Their History*. Cassell, Petter, and Galpin.

Schröter M, Van der Zanden EH, van Oudenhoven AP, Remme RP, Serna-Chavez HM, De Groot RS, and Opdam P (2014) Ecosystem Services as a Contested Concept: A Synthesis of Critique and Counter-Arguments. *Conservation Letters* 7(6), 514–523.

Schwitzgebel E (2013) A Dispositional Approach to Attitudes: Thinking Outside of the Belief Box. In Nottelmann N (ed.), *New Essays on Belief: Constitution, Content and Structure*. Springer, 75–99.

Seton JM and Seton ET (1930) *The Rhythm of the Redman: In Song, Dance and Decoration*. Ronald Press Co.

Shao W (2017) Weather, Climate, Politics, or God? Determinants of American Public Opinions toward Global Warming. *Environmental Politics* 26(1), 71–96.

Shkolnik A, Taylor CR, Finch V, and Borut A (1980) Why Do Bedouins Wear Black Robes in Hot Deserts? *Nature* 283(5745), 373–375.

Smith AB and Katz RW (2013) US Billion-Dollar Weather and Climate Disasters: Data Sources, Trends, Accuracy and Biases. *Natural Hazards* 67(2), 387–410.

Smith JE (1969) Time, Times, and the "Right Time": "Chronos" and "Kairos." *The Monist* 53(1), 1–13.

Somerville J (1941) Umbrellaology, or, Methodology in Social Science. *Philosophy of Science* 8(4), 557–566.

Spence CC (1961) A Brief History of Pluviculture. *The Pacific Northwest Quarterly* 52(4), 129–138.

Stemp WJ (2016) Explorations in Ancient Maya Blood-Letting: Experimentation and Microscopic Use-Wear Analysis of

Obsidian Blades. *Journal of Archaeological Science: Reports* 7, 368–378.

Stewart-Kroeker S (2022) Sacrifice in Environmental Ethics and Theology. *The Journal of Religion* 102(2), 237–261.

Strauss S and Orlove BS (2021) *Weather, Climate, Culture.* Routledge.

Thoreau HD (1881) *Early Spring in Massachusetts: From the Journal of Henry D. Thoreau.* Boston: Houghton, Mifflin and Company.

Vercler HR (1920) More about the "Rain-Maker." *Bulletin of the American Meteorological Society* 1(7–8), 80–82.

White AD (1896) *A History of the Warfare of Science with Theology in Christendom.* George Braziller.

Willoughby H, Jorgensen D, Black R, and Rosenthal S (1985) Project STORMFURY: A Scientific Chronicle 1962–1983. *Bulletin of the American Meteorological Society* 66(5), 505–514.

Wynne B (1992) Uncertainty and Environmental Learning: Reconceiving Science and Policy in the Preventive Paradigm. *Global Environmental Change* 2(2), 111–127.

INDEX

Abu Bakr al-Razi, 128
Abu Yusuf, 127
acequia communities, 98
Agobard of Lyon, 173
agricultural revolutions, 116
   first, 87–90
   fourth, 188–189
   second, 95–97
   third, 100–101
Alhazen, 128
Al-Jahiz, 127
Anderson, Paul Thomas, 31–32
aquaelicium, 169
Aristotle, 121–126, 170–173
astrolabe, 128
aurora borealis, 14, 130
autonomy, 201
Averroes, 128
Avicenna, 71, 73
Aztec tradition, 26, 167–169

Bacon, Francis, 130, 176
Badenov, Boris, 199–202
Bayes, Thomas, 153
Beck, Ulrich, 113–115
bomb cyclone, 62
Boulder floods, 40–42
Bradbury, Ray, 29
British agricultural revolution. *See* agricultural revolutions, second

Bulletin of the American Meteorological Society, 137

Camus, Albert, 30–31
Canon of Medicine, 74
Central Arizona Project, 99
Chaucer, 28, 128
Chimú civilization, 169
Chinese tradition, 26, 94, 120
Chronos, 22
climate, 18–20, 58, 75, 134, 138–140, 142–145, 155, 159
communication, 62
constructionism, 108–115
corn sweat, 9
crop insurance, 54

Darwin, Charles, 77, 136, 176
demiurge, 172
Democritus, 123
Dennett, Daniel, 183
Descartes, René, 131, 176
destruction, 45–52
determinism, 69–78, 123, 128, 134, 166
Diamond, Jared, 78
Doctor universalis. See Magnus, St. Albertus
dual use, 185–188
Dungeons & Dragons, 8

INDEX

Dust Bowl, The, 180
Dyn-O-Gel, 182

ecosystem services, 51, 63–69, 101–108, 113, 205
Egyptian gods, 26
Eliot, T. S., 29
ENMOD, 186–188
Environment and History, journal, 64
Epictetus, 85
etymology, 21–24
Euripides, 85
extraterrestrial weather, 13–15, 21–24

faint young Sun paradox, 12
Faulkner, William, 28
fire tornadoes, 6
forces, 37–39, 42–45, 86, 116–117, 124, 157–158, 169, 173, 198–202
Franklin, Benjamin, 133
French Revolution, 65
frost flowers, 7
Fu-Go balloon bombs, 184

Galen of Pergamum, 73
Galileo, 130
Genesis, 170
Giddens, Anthony, 113–115
Goldilocks Zone, 12
Green revolution. *See* agricultural revolutions, third

haboobs, 7
hair ice, 7

Hammer of Witches, The. *See Malleus Malleficarum*
Hatfield, Charles, 178
hazards, 97, 110, 113–115
Heat Miser, the, 65
Heidegger, Martin, 111–113
Helios, 165
Herodotus, 71
Hesiod, 74, 93, 121, 135
heteronomy, 201, 207
Hindu mythology, 26
Hippocrates, 71, 128
Homer, 25, 125–126
hucksters, 178–181
Hume, David, 76
humorism, 73
Hurston, Zora Neale, 29

Ibn al-Haytham. *See* Alhazen
Ibn Rushd. *See* Averroes
incommensurability, 57–59
indeterminacy, 142–145
Indra, 26
intentionalism, 178–188
interplanetary weather, 14

jökulhlaup, 6
Joon-ho, Bong, 31
Joyce, James, 30

Kairos, 21–23, 34, 115, 201–202
Kepler, Johannes, 130
Kneipp, Sebastian, 67
Kratos, 171, 201

INDEX

Lamarck, Jean-Baptiste, 77
Lebensraum, 77
Lee, Spike, 31
Leibniz, Gottfried Wilhelm, 133
Locke, John, 133, 176

Magnus, St. Albertus, 122
Maimonides, 75
Malleus Malleficarum, 174
mammatus clouds, 7
Marcus Aurelius, 85
Mayan tradition, 27, 167–169
mechanistic model, 175–178
megadrought, 62
metaphor, weather as, 33–36, 50
meteorophilately, 224
monotheism, 171–173
monsoons, 87, 90, 113, 185
Montesquieu, 76, 217
moral categories, 45–52
Myth of the Metals, 74
Mythology, 24–27

National Flood Insurance
    Program, 54
Native American tradition,
    167–168
Natural Cure Movement, 67
Neolithic revolution. *See*
    agricultural revolutions, first
Newton, Isaac, 133, 148

Occasio, 202
Odyssey, 25, 125, 168, 171
On Airs, Waters, Places, 72

Operation Commando Lava, 185
Operation Popeye, 185

Pascal, Blaise, 133, 153
pathetic fallacy, 33, 183
Perun, 25, 202
Phaeton, 165
philately, thematic, 224
piloi, 93
plant hardiness zones, 159
Plato, 74, 153, 172
plenum, Aristotelian, 124
Pliny the Elder, 126
pluviculture, 178–183
polytheism, 171–173
Pope Innocent VIII, 174
precisificiation, 151
production, 51
Project Stormfury, 182, 185
Protagoras, 62, 151
Ptolemy, Claudius, 125, 127
puquios aqueducts, 98

qanat, 98

rainbows, Aristotle on, 121
Ratzel, Friedrich, 77
Rehder, Alfred, 159
risk, 96–97, 113–115, 149,
    153–157, 179, 187–188, 206

sacrifice, 168–170
Seasonal Affective Disorder, 67–68
Shakespeare, 30, 174
snow, words for, 148

Social Determinants of Health, 55
Star Wars, 44
statistics, 149–157
Steinbeck, John, 29
Stoicism, 85
Swift, Jonathan, 29

teleological model, 170–175
telos, 170, 172, 174
Tempestarii, 173
Thales of Miletus, 122
Theophrastus of Eresus, 124
Thor, 26
Tolstoy, Leo, 30

torpor, 69
Twain, Mark, 19
Twin Earth, 15

USDA, 54, 101, 160

Weatherator 9000, 194–199
Whitehead, A. N., 154
wind, 72–73, 124–127
windmills, 99, 112

Year Without a Santa Claus, The, 65

Zeus, 22–26, 171, 202

For EU product safety concerns, contact us at Calle de José Abascal, 56–1°,
28003 Madrid, Spain or eugpsr@cambridge.org.

www.ingramcontent.com/pod-product-compliance
Lightning Source LLC
LaVergne TN
LVHW011812060526
838200LV00053B/3752